A Career-Focused Introduction to
NANOSCALE MATERIALS
TECHNOLOGY

TANIA M. CABRERA

A CAREER-FOCUSED INTRODUCTION TO NANOSCALE MATERIALS TECHNOLOGY

iUniverse books may be ordered through booksellers or by contacting:

iUniverse
1663 Liberty Drive
Bloomington, IN 47403
www.iuniverse.com
1-800-Authors (1-800-288-4677)

Because of the dynamic nature of the Internet, any web addresses or links contained in this book may have changed since publication and may no longer be valid. The views expressed in this work are solely those of the author and do not necessarily reflect the views of the publisher, and the publisher hereby disclaims any responsibility for them.

Any people depicted in stock imagery provided by Thinkstock are models, and such images are being used for illustrative purposes only.
Certain stock imagery © Thinkstock.

ISBN: 978-1-4917-8613-0 (sc)
ISBN: 978-1-4917-8612-3 (e)

Library of Congress Control Number: 2016901110

Print information available on the last page.

iUniverse rev. date: 2/4/2016

A Career-Focused Introduction to Nanoscale Materials Technology

Tania M. Cabrera
Associate Professor of Nanoscale Materials Technology
Division of Math, Science, Technology & Health
Schenectady County Community College

Art by Alicia Faucett, MS

Edited by Teresa Jacques, Ph.D.

Dedicated to My Students

"The secret in education lies
in respecting the student."

— *Ralph Waldo Emerson*

CONTENTS

Preface

I started my teaching career at a community college in 2009, when I was hired to develop and implement a Nanoscale Materials Technology program. I live in the Albany, New York region, which has since been nicknamed "Tech Valley," as we have seen a huge growth in the technology industries in recent years. My background did not necessarily fit the profile of a Nanoscale Materials Technology professor who specializes in semiconductor chip manufacturing. Upon high school graduation, I attended Simmons College in Boston, MA, where I double majored in Chemistry and Physics of Materials. Under the careful guidance of my organic chemistry professor and advisor, Dr. Rich Gurney, I became enamored with materials science research and completed a project doing microcontact printing on gold surfaces. I attended a summer internship program after my sophomore year at Simmons (National Science Foundation's Research Experience for Undergraduates). The program I got into was at Columbia University in the late Nicholas Turro's research group. It was during this time that I gained a real passion for nanotechnology, and the curious world that exists at the nanoscale. As a result, I returned to Columbia for graduate school, where I continued my work making functional nanoparticles.

After graduate school, I was hired by my amazing mentor Dr. Ruth McEvoy, former chair of the Math, Science and Technology department at Schenectady County Community College (SCCC). As I mentioned, the school was very forward thinking in developing a Nanoscale Materials Technology program to train technician level students for a new workforce in our area. Despite my lack if background in the area, I spent time studying the subjects I would need to be able to teach students relevant material. I learned a lot from local experts at a company called SuperPower, particularly the late Dr. Adrei Rar, who was always willing to give of his time to bring me up to speed on topics ranging from vacuum science to microscopy to thin film deposition.

As much as I loved teaching and the students at SCCC, I was faced with a somewhat difficult situation. I needed to teach students at a level for which no textbook existed. These students, for example, do not take calculus as traditional engineering students would, but needed more in depth information than could be found in general books about nanotechnology. It is for this reason that I am writing this book. Its goal is to inform at an appropriate level for community college students who are preparing for careers as technicians in the semiconductor manufacturing field.

My students are my inspiration. As such, you will notice a section at the end of each chapter called "NanoSpeak," where I interview some of my former students to give you a real-world feel for their opinions on the field. I asked them a series of questions and told them to answer as if they were speaking to prospective nanotechnology students. These are their exact words, and I hope you will draw from their insights. I also ask you some questions, or things to think about after you read each chapter. Keep thinking and don't be afraid of making inferences!

The book is formatted to begin with a general introduction to materials science, followed by specific insights into the world of nanoscale materials, and then a bit about careers and working in the industry. I hope you find it helpful, either as an educator, student, or curious party.

Author's Acknowledgements

I would like to thank Dr. Ruth McEvoy for taking a chance on an unexperienced teacher, and having faith in me to develop and shape Nanoscale Materials Technology education at Schenectady County Community College. Thank you for being an outstanding professional example, mentor and friend.

Thank you to my amazing colleague Syeda Munaim for assisting me with Chapter 8 and being a biology guru.

I would also like to thank Alicia Faucett for making all of the images in this textbook, and Teresa Jacques for spending countless hours editing every single page herein. I am blessed to have both of these women in my life.

Finally, thank you to my family who have helped me out in countless ways, and never fail to support me. I love you Miguel, Sofia, Magdalena, Babbo, Chuqui, Chachi, Bianca, Nelson and Oscar.

CHAPTER 1 AN INTRODUCTION – WHAT IS MATERIALS SCIENCE

KEY TERMS

MATERIALS	AMORPHOUS
PROPERTIES	CRYSTALLINE
STRUCTURE	ELASTOMER
PROCESSING	THERMOPLASTIC
PERFORMANCE	THERMOSET PLASTIC
METAL	CONCRETE
ALLOY	MATRIX
CERAMIC	BINDER
POLYMER	REINFORCEMENT
COMPOSITE MATERIAL	METAL MATRIX COMPOSITE
ADVANCED MATERIAL	CERAMIC MATRIX COMPOSITE
SUPERALLOY	REINFORCED PLASTIC
GLASS	SANDWICH STRUCTURE

INTRODUCTION

In this chapter you will be introduced to the field of materials science. You will learn some of the history of materials and their importance to human development, and gain some insight into the major classes of materials. A thorough understanding of materials science will help you to understand what makes nanoscale materials so unique and special. The relationship between the structure and performance of materials will be outlined for you in the introductory chapters of this text, which will then be compared to materials on the nanoscale. Many examples of applications of materials will be given to provide you with a real-world perspective of materials science and nanoscale materials.

Some things to think about:

- What is the difference between a chemical substance and a material?
- What are some everyday examples of useful metal, ceramic, and polymer materials?
- When have you benefitted from the use of a composite material?

1.1   HISTORY

Materials science is generally defined as the science describing the relationship between the structure and properties of materials. **Materials** can be defined as a substance from which other useful things can be made. The term **properties** refers to *both* physical and chemical properties. Properties include how a material reacts to such things as heat, magnetic or electric fields, physical force, or chemicals. The term **structure** refers to the arrangement individual atoms that make up the material. If we want to change any of a material's properties, we must do so by changing its structure. This is known as **processing**, or how we chemically change a material. Finally, we consider how the material can be used, or, its **performance**. A material's performance depends on its properties.

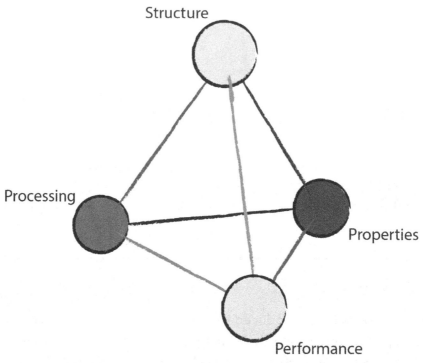

FIGURE 1.1

Figure 1.1 Relationship between structure, properties, processing and performance of materials

Though you may not think of it, the history of materials goes back many thousands of years. Historians have actually charted man's development by recognizing which materials were used to make objects such as tools and weapons, for example, the Stone, Bronze, and Iron Ages. The earliest humans used available naturally occurring materials, such as stone and wood. As time passed, man discovered that it was possible to produce materials that had superior properties to those that were naturally occurring. Heat could be used to soften and melt metals. Bronze, made from molten copper and tin, was the first alloy, or mixture of metals. Bronze is much easier to mold and shape than stone, making it a superior material.

Although we still use bronze today, after approximately 1000 B.C., iron ore obtained from the earth allowed man to work iron. There was no way to melt pure iron, but iron that contained a large percentage of carbon from charcoal could be more easily melted and cast into different shapes. The Iron Age continues today, although many new types of materials have been developed, such as ceramics, glasses, and semiconductors. We have the ability to design many new materials due to our study of materials, their structure, and properties.

1.2 CLASSES OF MATERIALS

We generally classify all solid materials into four main groups: metals, ceramics, polymers, and composites. The classification of metals, ceramics, and polymers is based on the types of atoms within the material, and how the atoms are bonded. There are also materials that are some combination of these three basic categories, which can be classified as **composites**. In addition to these categories, there are materials used for high-tech applications known as **advanced materials**. Semiconductors and nanoscale materials are examples of advanced materials.

METALS AND METAL ALLOYS

In the periodic table of the elements, the metallic elements are those that occur on the left-hand side of the chart. As of the year 2006, there are 117 known elements. Of those 117, 88 occur naturally, and of those 88, only 17 are non-metals (for example, oxygen, nitrogen, and chlorine), and 9 are semi-metals (for example, silicon, germanium, and arsenic).

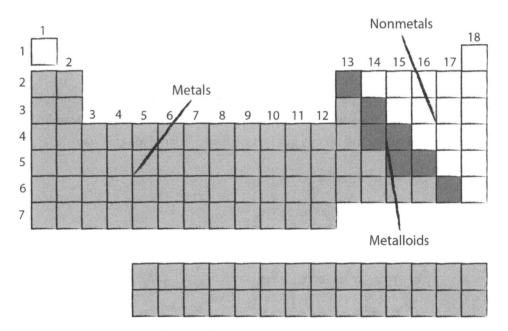

FIGURE 1.2

Figure 1.2 Location of Metallic elements on the periodic table

The atoms in a **metal** are generally arranged in a very ordered manner, and packed in quite densely. Metals may have physical properties such as ductility, resistance to fracture, opaqueness, luster, and ability to conduct heat and electricity. These physical properties are a direct result of a metal's atoms, and the electrons therein. This will be explained in chapter 2, section 2.6.1.

Materials that are classified as metals may also actually be combinations of metals, known as **alloys**. Common alloys include steel, brass, and bronze. Steel is a mixture of iron and carbon, brass is composed of the metals copper and zinc, while bronze is a mixture of copper and tin. **Superalloys** are alloys that exhibit high mechanical strength, good surface stability, and resistance to corrosion and oxidation at high temperatures. Superalloys are generally made using a base of nikel or cobalt.

CERAMICS AND GLASSES

A **ceramic** can be described as a compound that occurs between a metallic element, and a non-metallic element (usually oxygen, nitrogen or carbon). When we think of the word "ceramic," the image of a clay pot made in art class may come to mind, or perhaps the tiles on our kitchen floor. These "traditional ceramics" are made from what we call clay, a generic term to describe metal oxides often mixed with organic carbon-based materials.

The physical properties of ceramics include stiffness, strength, and the relative inability to conduct heat and electricity. In addition, ceramics are generally very brittle and can fracture easily. Ceramics can be quite useful since they can withstand very high temperatures without melting. These properties arise from the type of bonding that occurs in ceramics, which leads to highly ordered crystalline solids. Further explanation of these bonds can be found in sections 2.6.2 and 2.6.4,

Glasses are a special type of ceramic. While glasses are generally a metal or semimetal bonded to a non-metal, they do not have a crystalline structure like a ceramic. Instead, they are **amorphous**, meaning the atoms are not ordered in any special way.

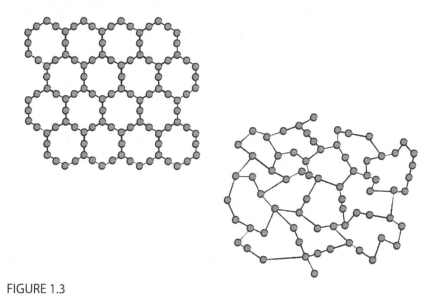

FIGURE 1.3

Figure 1.3 Crystalline vs. Amorphous structures

POLYMERS

The term **polymer** comes from the Greek "poly," which means many, and "meros," which means parts. As such, we can think of polymers as very long molecules that are composed of many repeating parts. A good image to represent such a molecule would be a long chain of paper clips hooked together.

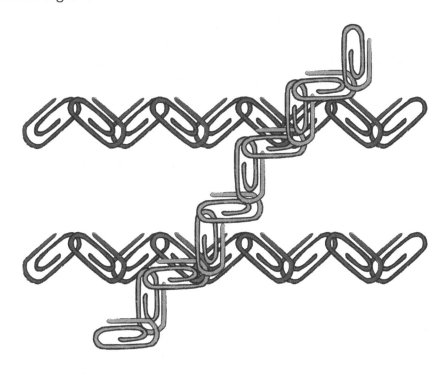

FIGURE 1.4

Figure 1.4 Paper clip representation of a polymer material

Thousands of different polymers exist, but we are generally most familiar with those that make up common plastic goods such as drink bottles. Most polymers are carbon- or silicon-based molecules bonded to other non-metals such as hydrogen, oxygen, or nitrogen. The chemical reactivity of polymers is typically quite low, and they can have a wide range of physical properties. In general, polymers are soft, pliable, and easily melted. They also do not conduct heat or electricity well.

Some common categories of polymers are:

- Elastomers
- Thermoplastics
- Thermoset plastics

Thermoplastics become moldable or pliable when warmed to a critical temperature specific to each polymer. They are different from **thermoset plastics** in that thermoset plastics undergo an irreversible chemical change upon heating, referred to as 'curing.' **Elastomers** are often called 'rubber,' and are characteristically very pliable.

COMPOSITES

The term composite comes from **composite material**, meaning a material that is made up of two or more materials with different physical properties. Once the different materials are combined into a composite, the result is a new material with unique physical properties from the individual starting materials. The materials in a composite do not undergo a chemical reaction with one another when they are joined. Instead, they maintain their chemical character. Generally, a new composite material is engineered to have desirable physical properties such as being lighter, stronger, and less expensive than the original starting materials.

The two main components of a composite material are the bulk of the material known as the **matrix**, or **binder**, and the **reinforcement,** smaller fragments of very strong material encompassed by the matrix.

SUMMARY OF TYPES OF MATERIALS:

METALS AND METAL ALLOYS

- Iron
- Carbon steels
- Stainless steel
- Aluminum
- Copper
- Magnesium
- Nickel
- Titanium
- Precious metals (platinum, palladium)
- Refractory metals (tungsten, molybdenum)
- Superalloys (Hastelloy, Inconel)

POLYMERS

- Thermoplastics
- Elastomers
- Thermoset plastics

CERAMICS AND GLASSES

- Glasses
- Glass ceramics
- Graphite*
- Diamond*

COMPOSITES

- Reinforced plastics
- Metal matrix composites
- Sandwich structures
- Concrete
- Ceramic matrix composites

*Graphite and diamond are composed of only carbon, and as such are not typical ceramics. They are classified as such based on their physical properties.

Some of the earliest composite materials were bricks made of combining straw with mud as far back as 6,000 years. Other very common composites we still use are plywood, or wood glued together at various angles, and **concrete** which is a mixture of small, coarse particles embedded in cement.

Newer composite materials include **metal** and **ceramic matrix composites**, **reinforced plastics**, and **sandwich structures**. As the names would suggest, in a metal matrix composite, a metal serves as the binder material with non-metal reinforcements, and in a ceramic matrix material, the bulk of the material is ceramic. Reinforced plastics are composites wherein a polymer serves as the matrix material. A common reinforcement in a polymer matrix is carbon fiber to add strength to a plastic without making it heavy. A sandwich composite is a material composed of two thin, sturdy skins that are attached to a lightweight, thick core of lower strength, flexible material. The resulting composite is low density and bendable yet stiff.

NANOSPEAK - WITH MATTHEW HUDMAN

1. Where do you work and what is your job title?
    GE Global research, Fab operations technician

2. What is your educational background?
    AAS Automotive Technologies, AAS Nanoscale Materials Technology from Schenectady County Community College.

3. How did you get interested in nanotechnology/where did you hear about it?
    I got interested in nanotech when an injury prevented me from returning to my previous career as an auto technician. I began researching the different programs offered through the community colleges in the area and found the nanoscale program very intriguing.

4. What advice would you give someone who might want to get into the field?
    Collaboration and teamwork environments are not going away. Learn to work with others, learn how to present your ideas to others, learn how to communicate effectively. Have a positive attitude towards challenges and have respect for others. One other thing that I have seen is that integrity and knowing when to ask for help is a trait that is highly desirable. The education and training tells the employer that you are teachable. Everything else is just as important if not more so.

CHAPTER 1 SUMMARY:

- Materials are substances out of which useful things can be made.
- Materials science is an interdisciplinary branch of science and engineering that examines the properties, performance, and processing of materials.
- Historians chart human development based on which materials were used for items such as tools and weapons.
- The main classes of materials are metals, ceramics, polymers, and alloys.
- Metals are materials composed of metallic elements which are found on the left-hand side of the periodic table.
- Metals are generally tough, ductile, dense, and good conductors of heat and electricity.
- Alloys are made of specific mixtures of metals.
- A ceramic is a material made of a metallic element bonded to a non-metallic element (found on the right-hand side of the periodic table).
- Ceramics are stiff, brittle, are able to withstand high temperatures and are electrical insulators.
- Polymers are long molecules made of repeating carbon or silicon units.
- We commonly think of polymers as "plastics." They are usually chemically inert, soft, moldable, and cannot conduct heat or electricity.
- Composites are combinations of two or more materials that benefit from properties of each.
- We generally are trying to achieve strong and lightweight properties when combining materials to make a composite.

REFERENCES

1. W. Callister and D. Rethwisch, *Materials Science and Engineering, An Introduction, 9th Ed.,* John Wiley & Sons, 2014.
2. P. Thrower, *Materials in Today's World, 2nd Ed.,* Learning Solutions, 1995.

CHAPTER 2   ATOMS AND BONDING

INTRODUCTION

We will begin this chapter by learning about the structure of atoms. The history of the atomic structure of matter will be described, as well as some very important experiments that led to the understanding that matter is made up of tiny particles called atoms. The way which elements are arranged into a periodic table will be illustrated, followed by an understanding of the importance of subatomic particles called electrons.

It is necessary to understand what an atom looks like in order to learn about the impetus for atoms bonding and the different types of chemical bonds that exist. In the next part of Chapter 2, we will outline the types of chemical bonds that exist and how these chemical bonds provide a material's overall structure and in turn dictate its physical properties. Once a material's physical properties are known, its use can be determined.

Some things to think about:

- Which scientists made major contributions to our current understanding of atomic structure?
- Which types of subatomic particles exist in atoms?
- What is so special about the arrangement of elements in the periodic table?

- What is the reason for atoms to undergo chemical bonding?
- How is a metallic bond different from an ionic bond and a covalent bond?
- How are secondary bonds different from primary bonds?
- What are some examples of secondary bonds having an impact on a material's properties?

2.1   ATOMIC THEORY OF MATTER

The idea that matter is made of small particles goes back many thousands of years. Around 460 B.C., the Greek philosopher Democritus posed this question: If you continually break matter in half, how many times would you have to break it until you could no further break it? He proposed that the endpoint would be a tiny bit of matter which he named the atom, coming from the Greek "atmos," which means indivisible.

Democritus' idea was not well-received by other philosophers at that time, and it was not until the 1800s that scientists again began asking the important question: How is matter structured?

In 1808, English scientist John Dalton came up with an explanation of the structure of matter, which he believed was made of combinations of very small particles. Dalton made four postulates, or, assumptions about matter:

1.   All matter is composed of indivisible atoms. An **atom** is an extremely small particle of matter that maintains its identity during a chemical reaction.
2.   An **element** is a type of matter composed of only one kind of atom. Each atom of a given kind has the same properties; for example, atoms of the same element have the same average atomic mass.
3.   A **compound** is a type of matter composed of two or more elements chemically combined in fixed portions.
4.   A **chemical reaction** is the rearrangement of the atoms present in the reacting substances to create new chemical combinations present in the products. Atoms are not created, destroyed, or altered during the reaction.

Although Dalton had hypothesized that the atom was indivisible, future experiments would show that atoms themselves are made up of smaller particles. He also provided no real explanation of what the atom looks like; this was determined in the years following Dalton's initial atomic theory.

2.2   STRUCTURE OF THE ATOM

In 1897, British physicist J.J. Thomson conducted a series of experiments proving that atoms were not indivisible particles. The apparatus used by Thomson included two electrodes from a high-voltage source sealed in a vacuum-pumped glass tube. The negatively charged electrode is known as a *cathode* and the positive electrode is called an *anode*. The tube also contained positive and negative electrically charged plates. When the high-voltage current was turned on, the glass tube emitted a green light caused by the interaction of the glass with rays originating from the cathode (*cathode rays*).

After the cathode rays exited left the negative electrode, they moved toward the anode where some rays passed through a hole to form a beam. The beam bent away from the negatively charged plate and toward the positively charged plate.

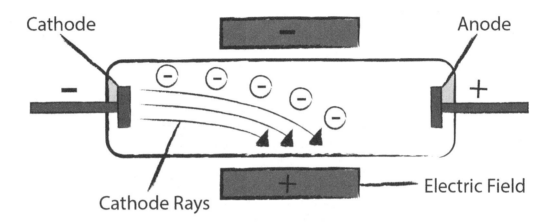

FIGURE 2.1

Figure 2.1 Formation of cathode rays

Thomson showed that the characteristics of cathode rays are independent of the material making up the cathode. From this evidence, Thomson concluded that a cathode ray consists of a beam of negatively charged particles he called **electrons**, and that electrons are constituents of all matter.

To further support Thomson's findings, in 1909 U.S. physicist Robert Millikan determined the actual numerical value for the charge on an electron through a series of experiments that involved observing how a charged drop of oil fell in the absence and presence of an electric field. In Millikan's oil drop experiments, an atomizer introduced a fine mist of oil droplets into the top of a chamber. Several drops fell through a small hole in a positively charged plate into the lower chamber, where the observer would watch the motion of the oil drop though a microscope. Some of the drops would pick up one or more electrons from the friction of the atomizer and became negatively charged. A negatively charged oil drop was attracted upward toward the positively charged plate. The drop's upward speed could be measured, and when related to an electron's mass-to-charge ratio, the charge and mass of the electron was obtained, determined to be 1.602 x 10^{-19} coulombs, and 9.109 x 10^{-31} kg, respectively.

Another scientist who worked on solving the mysteries of atomic structure was Irish-born mathematical physicist William Thomson, better known as Lord Kelvin. In 1910, Lord Kelvin thought that the atom might look something like plum pudding. He believed that the atom was a consistently positively charged cloud, with the electrons randomly scattered throughout, like the raisins in plum pudding. If you are not familiar with plum pudding, you may better relate to chocolate chip cookie dough. Imagine that the dough is a uniform positive charge, and the chocolate chips are electrons randomly distributed throughout the dough.

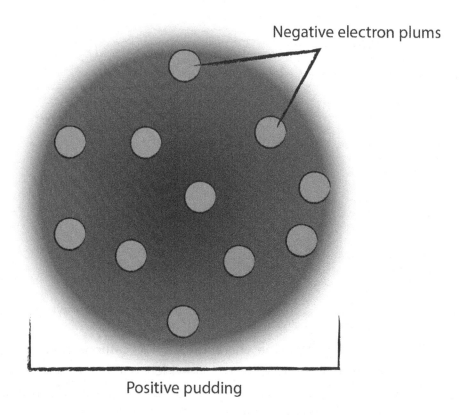

Negative electron plums

Positive pudding

FIGURE 2.2

Figure 2.2 Plum pudding model of the atom

In 1911, British physicist Ernest Rutherford dramatically changed our ideas about structure of the atom when his famous gold foil experiment led to the nuclear model of the atom. In this gold foil experiment, a thin piece of gold was bombarded with positively charged alpha particles that came from radioactive materials (for example, uranium). Most of the alpha particles passed right through the metal foil, but some (1 in 8,000) scattered at very large angles, sometimes almost backwards. From these findings, Rutherford proposed a model of the atom wherein most of the atom's mass was concentrated in a positively charged center that he named the **nucleus**, around which the negatively charged electrons move. While most of the mass of the atom is in the nucleus, it only occupies a small space in the atom.

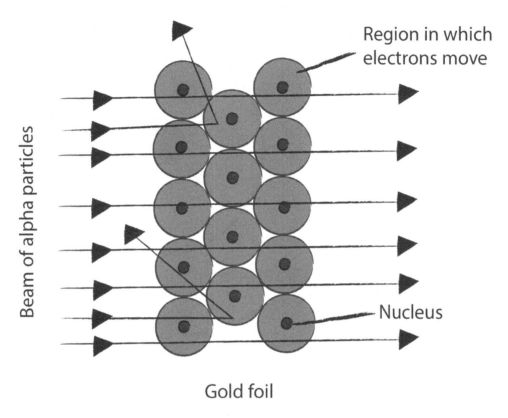

FIGURE 2.3

Figure 2.3 Rutherford's Gold Foil experiment

2.3   NUCLEAR STRUCTURE

Soon after Rutherford discovered the nuclear model of the atom, it was determined that the nucleus of the atom is composed to two different types of particles: protons and neutrons. Similar alpha particle-scattering experiments were performed by Rutherford, who discovered protons in 1919, and another British physicist, James Chadwick, who discovered neutrons in 1932. A **proton** is a nuclear particle with a positive charge equal to that of the electron, but a mass 1,800 times greater than an electron. A **neutron** is a nuclear particle with a mass similar to that of a proton, but carries no charge.

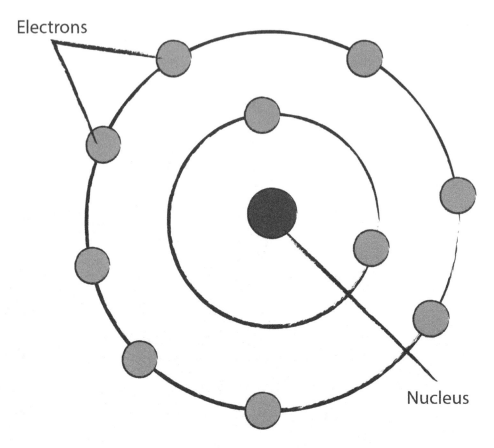

FIGURE 2.4

Figure 2.4 Early nuclear model of the atom

A **nuclide** is an atom characterized by a definite atomic number and mass number, and is denoted using nuclide symbols of the form . We can characterize a nucleus by its **atomic number** "Z," which is the total number of protons in the atom's nucleus. An element is now be defined as a substance in which all have the same atomic number. For example, if an atom has 5 protons in its nucleus, it is always a boron atom. The **mass number** "A" is the total number of protons and neutrons in a nucleus. For example, the nucleus of 99% of naturally occurring carbon atoms has 6 protons and 6 neutrons, meaning it has an atomic number Z=6, and an atomic mass A=12 (6 protons + 6 neutrons) and can be described using the nuclide symbol .

We know that an atom is normally electrically neutral, so that means it contains as many electrons (negatively charged) as protons (positively charged). In a neutral atom, the number of electrons is the same as the number of protons. All nuclei of a specific element has the same atomic number. They can, however, contain a different number of neutrons, giving the atom a different atomic mass. **Isotopes** are atoms in which the nuclei have the same atomic number but different number of neutrons. Going back to our example of carbon, there is a naturally occurring isotope with 6 protons and 7 neutrons yielding the symbol .

2.4   PERIODIC TABLE OF THE ELEMENTS

In 1869, two scientists working independently, Russian chemist Dmitri Mendeleev and German chemist J. Lothar Meyer, made a similar discovery. They found that if the elements were arranged in horizontal rows in order of atomic weight, the elements in each vertical column had similar properties. This arrangement of the elements in rows and columns is called the **periodic table**. Eventually, it was determined that arranging the elements by atomic number, rather than atomic weight, eliminated any discrepancies in property patterns, and thus, our modern periodic table (with the elements arranged by atomic number) was created. The basic structure of the periodic table is its division into rows and columns. A **period** is the elements in any horizontal row of the periodic table. A **group** is the elements in any column of the periodic table. We mentioned that the elements in a group have similar properties. For example, the elements in group IA, also called the alkali metals, are all soft metals that react strongly with water.

The elements on the periodic table are divided by a thick line often called the "staircase" with metals on the left of the line and non-metals on the right. As discussed in Chapter 1, **metals** are substances that generally have a characteristic luster and are good conductors of heat and electricity. **Non-metals** are elements that do not exhibit the characteristics of a metal. Most of the non-metals are gases or brittle solids. Bromine is the only non-metal that is liquid at standard room temperature and pressure. The majority of the elements that border the staircase are called either **metalloids** or **semimetals**. These are elements that have both metallic and non-metallic properties. They tend to be poor conductors of electricity at room temperature, but when heated can be moderately good conductors.

2.5   ELECTRONIC STRUCTURE OF THE ATOM

For the purposes of this chapter, we will consider a simplified model of the atom that we can liken to the solar system. Think of the nucleus as the sun at the center of the solar system, and the electrons going around the nucleus as planets orbit the sun in fixed orbits. These orbits are called **electron shells**, and the further away from the nucleus, the larger and higher energy the shell. Higher numbered shells can also hold more electrons.

The first shell, which is closest to the nucleus, can accommodate up to 2 electrons. The second shell can contain up to 8 electrons, and the third can hold up to 18. A simple formula can tell us how many electrons each shell can hold – $2n^2$, where n is the number of the shell. So, for example, the 5th shell can hold a maximum of 50 electrons (2×5^2).

We will concern ourselves with the first 8 electrons in a given shell (except the first, which can only hold two). The last group on the periodic table, group VIII, is known as the **noble gases**. These gases do not react readily with any other element or compound. This is because they all have the maximum 8 electrons in their outermost, or **valence shell**, which results in the lowest energy configuration. **Figure 2.4** shows the example of neon, which has a full valence shell, also called an **octet.**

Atoms with a complete octet in their outermost shell are very stable. All atoms tend toward a stable noble gas electron configuration with filled valence shells, and may achieve this through a process known as **bonding**.

2.6   BONDING

We will consider three main types of bonds. These bonds determine the structure of a material, and in turn, the material's properties.

2.6.1   METALLIC BONDING

As we mentioned previously, the majority of elements on the periodic table are metals. The way in which atoms of metals are held together is known as **metallic bonding**. Let's take a look at an atom of sodium, element number 11.

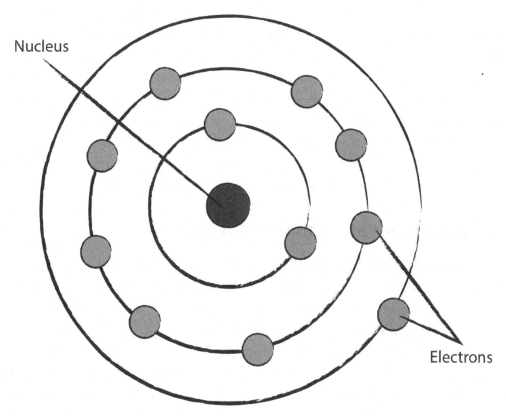

FIGURE 2.5

Figure 2.5 A sodium atom

Keep in mind - the goal of bonding is to achieve a valence shell that contains 8 electrons. As you can see, the sodium atom has a lone eleventh electron that takes up a third valence shell by itself. How can sodium make itself look like a noble gas? It can simply release that outermost electron. Afterward, the sodium atom contains 11 protons and 10 neutrons, giving it an overall positive charge. Charged atoms are called **ions**, and when the charge is positive, the ion is known as a **cation.**

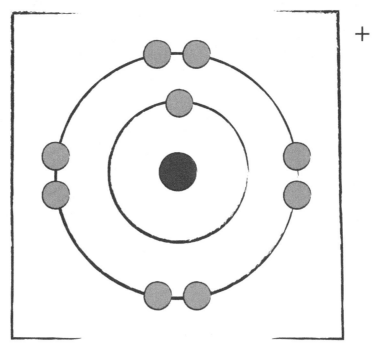

FIGURE 2.6

Figure 2.6 A sodium ion

The outermost electron released by a metal is called a "free electron." The metallic solid contains negatively charged free electrons, and positively charged metal cations. As we know from fundamental physics, opposite charges attract. It is this electrostatic attraction between free electrons and cations that holds metals together. To imagine this metallic bonding, think of the metal cations as marbles and the free electrons as glue that never fully dries holding the marbles together.

The free electrons in metallic bonding are always moving, and it is this property that allows metals to easily conduct heat and electricity. Notice in **Figure 2.7** that the metallic bond has no particular direction. If you go back to the idea of metallic boning as marbles held together by glue that never really dries, you can imagine that the marbles have the ability to roll over one another freely. This is why metals have the physical properties of malleability and ductility, or the ability to be hammered into thin sheets and drawn out into wires. You can also imagine that there is nothing to keep the marbles from packing in as tightly as possible next to each other. This is why metals are quite dense and heavy.

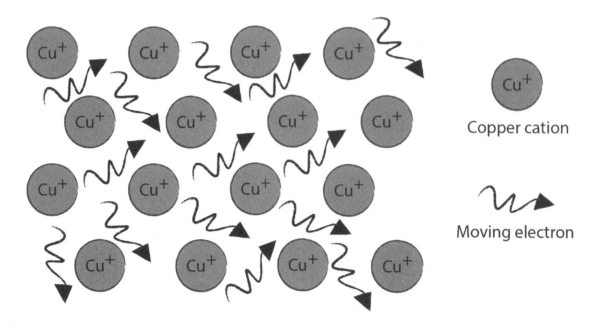

Copper cation

Moving electron

FIGURE 2.7

Figure 2.7 metallic bonding

2.6.2 IONIC BONDING

As noted in the discussion of metallic bonding, when an atom becomes charged by losing or gaining electrons, it is called an *ion*. Another way of achieving 8 electrons in an atom's outermost shell is via the loss and gain of electrons, a process called ionization.

To elaborate on this idea, go back to the image of sodium in Image 9. Again, sodium has an extra electron in a third electron shell that it wants to release to achieve a full outermost shell. Once the electron is lost, sodium has a positive charge, or is a cation, because it has 11 protons and only 10 electrons, leaving behind a +1 charge.

Now let's take a look at a chlorine atom, element number 17.

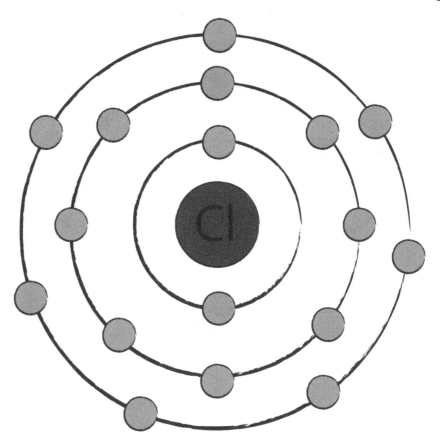

FIGURE 2.8

Figure 2.8 A chlorine atom

Notice that chlorine only has 7 electrons in its valence shell, leaving the need for one more to achieve an octet. Sodium has lost an electron to achieve a full valence shell, and chlorine will take that electron to fill its valence shell. Because chlorine has 17 protons and 18 electrons (after gaining one from sodium), it has an overall charge of -1. A negatively charged ion is known as an **anion**. Once again, we know that opposite charges attract, and this electrostatic force between the negatively charged anion and the positively charged cation is called an **ionic bond**.

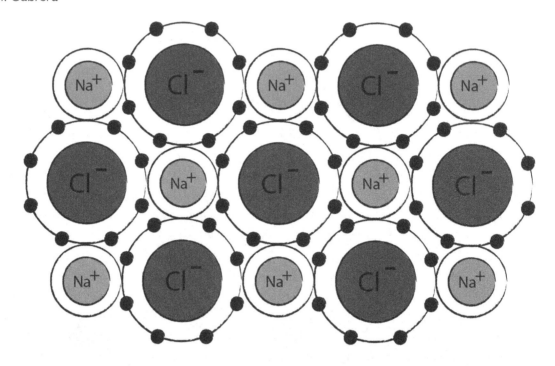

FIGURE 2.9

Figure 2.9 sodium chloride ionic bond

The example given here of sodium and chlorine is better known as common table salt. An ionic bond is so strong that we would have to heat table salt to 801°C, or 1,474°F, in order to melt it.

Figure 2.9 shows the arrangement of ions in table salt. Notice that, as in metallic bonding, the ions can attract each other in any direction, and therefore ionic bonds have no direction. As with metal cations in metallic bonding, the ions in ionic bonds will also pack together as closely as possible.

2.6.3   COVALENT BONDING

In the case of covalent bonding, two atoms will actually share electrons in order to achieve full outer shells. Take a look at a hydrogen atom, depicted in Image 14.

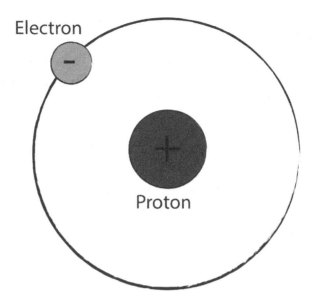

FIGURE 2.10

Figure 2.10 A hydrogen atom

As you can see, a hydrogen atom has one electron. The first electron shell needs two electrons to be fully stable and look like the nearest noble gas helium. In order to achieve full shells, the hydrogen atoms will overlap electron shells and share their electrons. These two shared electrons make up what is called a **covalent bond**. These two covalently bonded hydrogen atoms make up a hydrogen molecule, H_2. All naturally occurring hydrogen exists as H_2, since a single hydrogen atom with only one valence electron is so unstable.

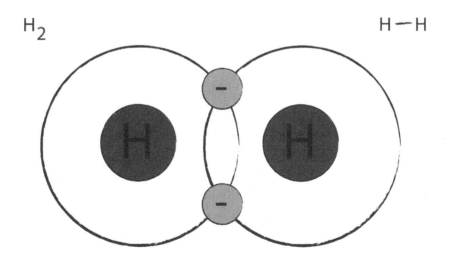

FIGURE 2.11

Figure 2.11 A molecule of hydrogen, H_2

Now let's consider an atom of carbon. Carbon is element number 6, and holds 2 electrons in the first shell, and 4 electrons in its outer shell. Diamond is simply made up of many carbon atoms covalently bonded together. The carbon atoms bond by sharing their 4 valence electrons with 4 neighboring carbon atoms. Each shared pair of electrons forms one covalent bond, so each carbon atom makes 4 covalent bonds.

In **Figure 2.12**, you can see the 4 shared electron pairs in diamond. Since electrons are all negatively charged, they will repel because like charges will repel. The bonds get pushed as far apart as possible, resulting in an angle of 109.5° between them creating a shape we call a *tetrahedron*. Image 16 illustrates the crystalline structure of the carbon atoms in diamond. Notice that the covalent bonds are **directional bonds**, in fixed positions in relation to the surrounding atoms.

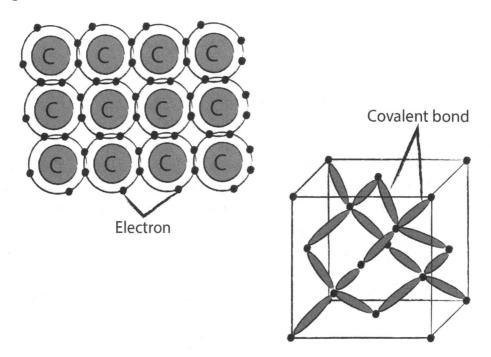

FIGURE 2.12

Figure 2.12 The carbon atoms in diamond

Since covalent bonds are directional, unlike metallic and ionic bonds, their atoms cannot pack in as closely, and, as a result, covalently bonded materials tend to be less dense. It is also difficult for these atoms to move past each other, so materials containing covalent bonds will not deform easily. Consider that diamond is the hardest naturally occurring material on earth. Since all of the valence electrons are shared in bonds, there are no free electrons (like there are in metals) available to conduct electricity, so covalently bonded materials are generally considered insulators.

2.6.4 POLAR COVALENT BONDS AND ELECTRONEGATIVITY

A covalent bond involves sharing at least one pair of electrons between two atoms. In the examples we gave of hydrogen molecules and diamond, the bonded electrons were shared equally. That is, the electrons spend the same amount of time around each atom. What happens when two atoms of different elements are bonded together? In these cases, the electrons are not necessarily shared equally, and the resulting bond is known as a **polar covalent bond**. A polar covalent bond can be defined as a covalent bond in which the bonding electrons spend more time around one atom than the other.

What determines which atom will pull the shared electrons more toward itself? The answer is a property of atoms called **electronegativity**. Electronegativity is the measure of the ability of an atom to draw electron density toward itself in a covalent bond. The most widely used electronegativity scale is that proposed by American-born chemist Linus Pauling.

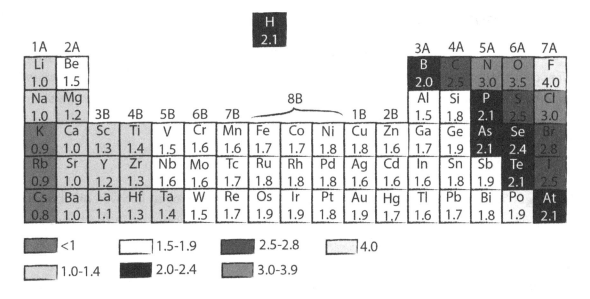

FIGURE 2.13

Figure 2.13 Pauling electronegativity scale

Elements are assigned an electronegativity value between 0.7 and 4.0. The value has no units. Fluorine is the most electronegative atom, meaning it has the ability to draw in electron density most strongly, and is assigned a value of 4.0. Cesium is the least electronegative atom and is assigned a value of 0.7. This makes sense considering that, in general, metals tend to lose electrons.

To determine how polar a bond is, we can simply determine the difference between the electronegativity values of the atoms. If the difference is 0, the electrons are being shared equally. If the difference is more than 0 but less than 1.9, there is an unequal sharing of electrons, therefore a polar covalent bond. If the difference in electronegativity is more than 1.9, the electrons are no longer being shared, and the bond is fully ionic.

Consider the bond between hydrogen and chlorine. Hydrogen has one atom in its outermost shell and requires one more to be full, and chlorine has 7 electrons in its valence shell, and also needs one more to complete its octet. Hydrogen and chlorine will share electrons; however, the electrons will not be shared equally between the two atoms. Chlorine has a high electronegativity value of 3.0, while hydrogen's electronegativity value is 2.1. Since 3.0 – 2.1 = 0.9, this is a polar covalent bond. Since chlorine has a higher electronegativity value, it will draw in electron density, and therefore the shared electrons will spend more time around chlorine than around hydrogen. This results is a partially negative charge on the chlorine atom, denoted δ-, and a partially positive charge around the hydrogen atom, denoted δ+.

2.7   SECONDARY BONDS

Metallic, ionic, and covalent bonds are all known as **primary bonds**, meaning they are directly involved in holding atoms together. There are also **secondary bonds**, or forces that exist between molecules. These secondary bonds often help dictate a material's strength, and are not as strong as primary bonds.

2.7.1   HYDROGEN BONDING

Hydrogen bonding is a secondary bond that exists when hydrogen is covalently bonded to a very electrogegative atom, such as, oxygen, nitrogen, or fluorine. As in the hydrogen and chlorine discussed in section 2.6.4, the resulting covalent bond is polar and has a partially positive and partially negative side.

Perhaps the most well-known example of hydrogen bonding is the secondary bond that exists between molecules of water. Water, which is made up of one oxygen and two hydrogen atoms, is a polar molecule, as illustrated in Image 18.

Because oxygen is very electronegative, it has the ability to pull electron density toward itself, inducing a partial negative charge. Since the shared electrons spend more time around oxygen than hydrogen, the hydrogen will have a partially positive charge. Now imagine putting two water molecules next to each other. Opposite charges attract, so the two molecules will orient themselves as shown in **Figure 2.14** with the negative end of one molecule attracted to the positive side of the second. This attraction between negative and positive is a hydrogen bond.

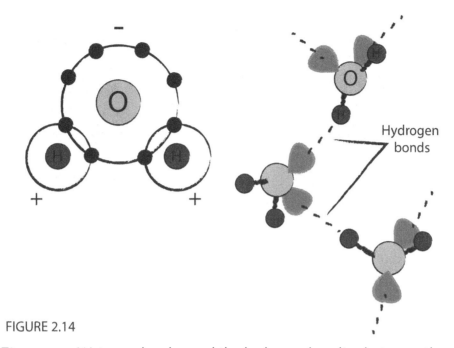

FIGURE 2.14

Figure 2.14 Water molecules and the hydrogen bonding between them

When several water molecules join together via hydrogen bonding at the correct temperature and pressure to form ice, they arrange themselves into hexagons. This hexagonal symmetry is the reason snowflakes have their distinct six-pointed-star shape. This same symmetry is why ice floats in water. Most liquids become denser when frozen into solid form. Water, however, becomes less dense due to the large space in the middle of the hexagon that results from hydrogen bonding.

2.7.2 VAN DER WAALS FORCES

Van der Waals forces, named for Dutch physicist J.D. van der Waals, are the weakest of all existing chemical bonds. Like hydrogen bonds, they are the result of attractions between partial positive and negative ends of a molecule, but are only temporary. To better understand these temporary forces that exist, let's consider two helium atoms. Helium, being a noble gas, has a full outer electron shell, is very stable, and generally does not undergo chemical reactions. The only bond that exists in helium is the van der Waals force that exists temporarily when the electrons are imbalanced. **Figure 2.15** shows that, at any one moment, the two electrons of one helium atom may arrange themselves on one side of the atom, creating a temporary partial negative on the side with the electrons, and a temporary partial positive on the side with no electrons. A very weak attraction can be formed between the atoms or molecules when this particular imbalance occurs.

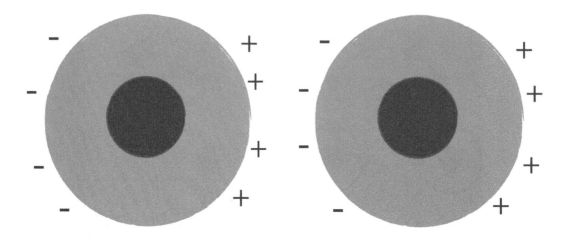

Van der Waals Interaction

FIGURE 2.15

Figure 2.15 Van der Waals forces

A good example of a material with both primary and secondary bonds dictating properties is graphite. Graphite is the familiar black solid found in pencil lead. In graphite, the carbon atoms are all bonded to only three others in two-dimensional sheets with a hexagonal shape. The sheets stack atop one another. Carbon has four valence electrons, so one un-bonded electron moves freely between the sheets. While the sheets contain carbon atoms strongly covalently bonded to one another, there are only weak van der Waals bonds between the sheets. For this reason, the sheets can easily slide over one another. This is why you can write on paper with graphite, and why it is often used as a lubricant.

Another example of primary and secondary bonds dictating the properties of materials is polymers. Polymers, as mentioned in Chapter 1, are long-chained molecules generally made up of carbon or silicon. The chains of atoms are covalently bonded together into long molecules that you can think of as like strands of spaghetti. The bonds holding the actual chain together are very strong. In between the strands, however, are only weak van der Waals bonds. These weak bonds between the strands allow them to easily move over one another. This structure is what causes most polymers to very plastic, or easily deformed.

NANOSPEAK – WITH ALEXANDER MCKENNA (CHAPTER 2)

1. Where do you work and what is your job title?
    I am a technician at SuperPower, Inc.

2. What is your educational background?

   I am currently working on finishing my AAS in a Nanoscale Materials Technology program.

3. How did you get interested in nanotechnology/where did you hear about it?

   I heard about the nanotechnology program when I was signing up for classes at my local community college. Having always known I wanted to be a part of the science community, I decided that this was a new field and something I knew very little about so I wanted to give it a try.

4. What advice would you give to someone who might want to get into the field?

   STUDY! And pay attention in your lab classes. At least where I work we have a lot of record keeping as I'm sure most places do. Being organized and efficient will get you to the top. And also be patient, even after the schooling there is a lot of information to take in when you start at one of these places. So pay attention and don't worry yourself too much, if you try to stay organized and pay good attention you will do great!

CHAPTER 2 SUMMARY

- John Dalton was the first modern scientist to put forward an atomic theory of matter wherein he stated all matter is made up of tiny particles called atoms.
- J.J. Thomson expanded on the atomic theory by discovering negatively charged subatomic particles, called "electrons," via his study of cathode rays.
- Robert Millikan used the oil drop experiment to determine the mathematical charge of electrons discovered by Thomson.
- Ernest Rutherford developed the nuclear model of the atom, wherein the majority of an atom's mass is concentrated in a central, positively charged nucleus.
- Rutherford later determined the nucleus contains positively charged subatomic particles called protons. His student James Chadwick discovered the nucleus also contains neutral subatomic particles called neutrons.
- We can characterize an atom's nucleus by its atomic number "Z," or the total number of protons in the nucleus.
- The mass number "A" of an atom refers to the total mass, or number of protons plus number of neutrons in an atom's nucleus.
- An isotope is an atom with the same number of protons but different number of neutrons.
- The periodic table of the elements was created by Dmitri Mendeelev and arranges the elements in order of increasing atomic number.
- The horizontal rows on the periodic table are called periods and the vertical columns are called groups. Elements in the same group have similar properties.

- Metals are elements that exist on the left-hand side of the periodic table, non-metals are on the right-hand side, and semimetals exist on the periodic table "staircase."
- Electrons move around the nucleus in fixed energy levels called electron shells.
- The outermost electron shell is known as the valence shell.
- Noble gases have full valence shells, which is the ideal, most stable configuration with the lowest energy.
- All atoms bond to achieve full valence shells.
- Primary bonds are forces that hold atoms together and may be metallic, ionic, or covalent.
- In metallic bonds, metal cations are held together by freely flowing electrons.
- Ionic bonds occur between a metal and a non-metal, when the metallic atom gives electrons to the non-metallic atom.
- Covalent bonds form when atoms share electrons to fill their valence shells.
- Polar covalent bonds exist when two atoms of different electronegativities bond and share electrons unequally.
- Secondary bonds do not hold atoms together.
- Hydrogen bonding is a secondary force that exists when hydrogen is bonded to a highly electronegative atom such as oxygen, nitrogen, or fluorine.
- Van der Waals forces exist when electrons temporarily become imbalanced and create instantaneous partial charges.

REFERENCES

1. S. Zumdahl and D. Decoste, "*Introductory Chemistry: A Foundation,* 7th Ed., Brooks Cole, 2010.
2. D. Ebbing and S. Gammon, "*General Chemistry,* 7th Ed., Houghton Mifflin College Div., 2002.
3. P. Thrower, *Materials in Today's World,* 2nd Ed., Learning Solutions, 1995.

CHAPTER 3   PHYSICAL PROPERTIES OF MATERIALS

KEYWORDS

SOLID

LIQUID

GAS

MELTING

FREEZING

VAPORIZATION

SUBLIMATION

CONDENSATION

DEPOSITION

PHASE DIAGRAM

EQUILIBRIUM

TRIPLE POINT

EXCITED ELECTRONS

GROUND STATE

BULK MATTER

REFRACTION

REFRACTIVE INDEX

LIGHT INTERFERENCE

CONSTRUCTIVE INTERFERENCE

DESTRUCTIVE INTERFFERENCE

DIFFRACTION GRATING

LIQUID CRYSTAL

TENSION TEST

ENGINEERING STRESS

ENGINEERING STRAIN

MODULUS OF ELASTICITY

ELASTIC DEFORMATION

PLASTIC DEFORMATION

TENSILE STRENGTH

DUCTILITY

TOUGHNESS

HEAT CAPACITY

TRAVELING LATTICE WAVES

PHONON

THERMAL EXPANSION

THERMAL CONDUCTIVITY

THERMAL STRESS

WAVE

WAVELENGTH

FREQUENCY PHOTONS

HARDNESS

ELECTRICAL CONDUCTION

ELECTRICAL CONDUCTIVITY

CONDUCTOR

INSULATOR

SEMICONDUCTOR

EMERGY BANDS

HIGHEST OCCUPIED MOLECULAR
   ORBITAL

LOWEST UNOCCUPIED MOLECULAR
   ORBITAL

ENERGY BAND DIAGRAM

FILLED/VALENCE BAND

EMPTY/CONDUCTION BAND

BAND GAP

INTRINSIC AND EXTRINSIC
   SEMICONDUCTORS

HOLE

P-TYPE AND N-TYPE

DOPING

DOPANT

PARAMAGNETIC

ANTIFERROMAGNETIC

FERROMAGNETIC

INTRODUCTION

In this chapter, we will expand upon the understanding of atoms and bonding gained from Chapter 2. A number of physical properties of materials will be addressed, along with a grasp of how those physical properties are a direct result of chemical bonding. The term 'thermal properties' refers to the way in which materials respond when heat is either added to or removed from them. The optical properties of materials, or how they interact with light, can come from a number of sources. Mechanical properties of materials describe the way in which materials behave when they experience an applied force. Electrical properties, or how materials respond to an applied electric field, will be defined, as well as magnetic properties, or how materials respond when exposed to a magnetic field. All of these physical directly relate to performance, or how the materials can be used.

Some things to think about:
- What is the difference between a material in the solid state versus the liquid state and the gaseous states?
- What needs to happen to a material in order to undergo various phase transitions?
- Which material has a higher heat capacity: a metal, ceramic, or polymer?
- Why do polymers have high linear coefficients of thermal expansion?
- Why do some ceramics conduct thermal heat via the phonon mechanism while metals conduct via the free electron mechanism?
- What are some of the phenomena by which we see materials as having various colors?
- If Young's modulus is defined as *stress divided by strain*, would polymers have a high or low value of Young's modulus?
- How are the band gap structures of metals, insulators, and semiconductors different?
- Why are most materials NOT magnetic?

3.1   THERMAL PROPERTIES

The term thermal property is meant to indicate how a material responds to changes in temperature.

3.1.1   STATES OF MATTER

Most commonly, a given material can exist in three different physical states, depending on temperature and pressure. Water, for example, can exist as a solid in the form of ice, liquid in the form of liquid water, and a gas in the form of steam. A **solid** can be defined as the state of matter characterized by rigidity, incompressibility, and having a fixed volume and shape. The molecules in a solid material are packed tightly in close contact with one another, and have no room to move around. The only movement molecules of a solid material can carry out are simple vibrations.

When looking at molecules in a **liquid,** we can see that though they are still packed in relatively tightly, they do have enough room to roll over one another. A liquid is also relatively incompressible like a solid and has a fixed volume, but unlike a solid, has no fixed shape. A liquid is pourable because there is enough room for the molecules to move over one another, and will take on the shape of its container.

Gases, on the other hand, are very easily compressible, and have no fixed volume or shape. A given volume of gas will fit into a container of almost any size and shape. The molecules in a gas are in constant, random motion. The molecules of a gas move in straight lines, in all directions, and at various speeds.

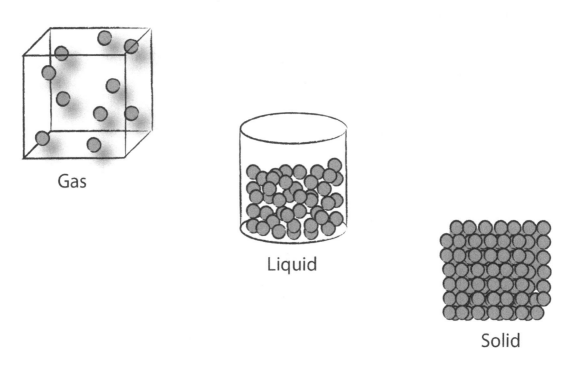

Gas

Liquid

Solid

FIGURE 3.1

Figure 3.1 Solid, liquid and gaseous phases of matter

3.1.2 PHASE TRANSITIONS

In general, the three states of matter have the ability to change into the other states via various processes that we will define in this section. **Melting** is the change of a material from the solid state to the liquid state. Sometimes, melting is also referred to as "fusion." An example of melting is ice or snow becoming liquid water. **Freezing** is the change of a material from liquid to solid state, and is the opposite of melting. Filling your ice cube tray with water and putting in into your freezer to make ice cubes is an example of freezing. Changing a material from a liquid to gas is known as **vaporization.** The change of a solid directly to a

Tania M. Cabrera

gas is called **sublimation.** There are a few solids that readily vaporize, for example, iodine and carbon dioxide (dry ice). Finally, gases undergo **condensation** or **liquefaction** when they change to the liquid state, and **deposition** when turned directly to a solid.

As mentioned at the beginning of the chapter, the solid, liquid, and gaseous phase of a material will exist at different temperatures and pressures. A phase diagram is a graph that helps us to summarize the conditions under which a material is solid, liquid, or gas.

Figure 3.2 is the **phase diagram** for water. It consists of three curves that divide the diagram into different regions labeled solid, liquid, and gas. In each region, the labeled state is the stable one for water at a given temperature and pressure. Every single point on a curve is a temperature and pressure at which the two phases labeled on either side are in **equilibrium**. Equilibrium is defined as the state when the rates of forward and reverse reactions are equal.

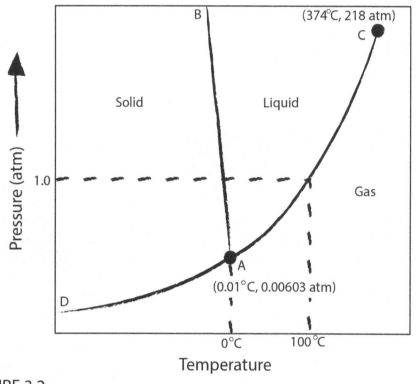

FIGURE 3.2

Figure 3.2 Phase diagram of water

The curve AB is the melting point curve. Since the melting point of water is not very much affected by pressure, you can see the curve is almost vertical. The melting point in the case of water decreases with pressure because liquid water is denser than ice. For most other substances, their solid state is denser than the liquid state, and so the melting point curve will lean slightly to the right.

The curve AC gives the vapor pressure of a material at various temperatures, as well as the boiling point for various pressures. For example, the boiling point of water at 1 atm, or standard pressure, is shown to be 100°C. The curve AD shows you the vapor pressure of solids at various temperatures. If you notice, all of the lines intersect at point A. Point A is known as the **triple point**, or the point at which all three phases of matter exist in equilibrium. For water, the triple point is 0.01°C and 0.00603 atm.

3.1.3   HEAT CAPACITY

When a solid material is heated, it absorbs some energy and experiences an increase in temperature. A material's ability to absorb heat from its external surroundings in known as its **heat capacity.** The heat capacity for a material represents the amount of energy that is needed to produce a specific rise in temperature. In mathematical terms, heat capacity (C) is expressed as change in energy (dQ) required to produce a temperature change (dT), shown in Equation 3.1 If you have already taken a course in chemistry, you may have also heard the term "specific heat" used to define heat capacity.

Equation 3.1 $C = \frac{dQ}{dT}$

In most solids, the main way in which thermal energy is absorbed is via an increase in the vibrational energy of the atoms. The atoms in a solid are always moving by vibrating, and each atom's vibrations affect the other atoms to which it is bonded. In other words, each atom's vibrations cause adjacent atoms to vibrate. When a solid is heated up, these coordinated vibrations produce what are called **travelling lattice waves**. You can think of these waves as sound waves that travel through a solid at the speed of sound. These waves can only have specific frequencies in the same way electrons in an atom can only exist in certain energy levels. A level of vibrational energy in a material is called a **phonon**.

3.1.4   THERMAL EXPANSION

Most materials will expand when you heat them, and contract when you cool them. Mathematically, the change in the length of a material with temperature can be represented in Equation 3.2 as

Equation 3.2: $\frac{\Delta l}{l_0} = \alpha_l \Delta T$

where l_0 is the initial length of the material, ·l is the change in length of the material (final length – initial length), ·T is the change in temperature (final temperature – initial temperature), and α_l is a parameter known as the linear coefficient of thermal expansion. α_l is a property of a material that will indicate the extent to which the material can expand upon heating.

If you think about thermal expansion from an atomic perspective, the bonds between the atoms will need to stretch. We know that metallic bonds are of intermediate strength since they are stronger than van der Waals bonds, but not as strong as covalent bonds. It makes sense, then, that the α_l of metals lie between that of polymers and ceramics. Ceramics are materials that contain very strong polar covalent or ionic bonds between metals and non-metals. Since the bonds are so strong, they do not stretch much, and therefore ceramics tend not to expand well upon heating. Of all the materials, ceramics have the lowest value of α_l. Polymers, on the other hand, experience a lot of weak van der Waals bonds between the chains, and have very high α_l values as a result.

3.1.5   THERMAL CONDUCTIVITY

Thermal conductivity is the property of a material that defines how well it can transport heat from high temperature regions to low temperature regions. There are two ways in which a material can transport heat. One of them is via the lattice vibrations, or phonons (3.1.3). The other method for conducting heat in a material is by free electrons. Free electrons in a hot region of a sample will have kinetic energy that they can transfer to atoms in a cooler area.

We know that pure metals' atoms are held together via metallic bonds, which are metal cations attracted by freely flowing electrons. Since metals contain these free electrons, the electron mechanism of transporting heat is much more efficient than the phonon mechanism. Ceramics and polymers, on the other hand, contain strong covalent bonds, and lack free electrons. In both ceramics and polymers, the main mechanism of thermal conductivity is via phonons. As such, both classes of materials are generally very poor conductors of heat, and in fact are often used as insulators. Think, for example, of the foam cup that holds hot coffee. It is polystyrene (STYROFOAM™), a polymer made of a styrene molecule backbone, and is used as an insulating cup to keep the coffee warm.

3.1.6   THERMAL STRESS

Thermal stress is stress induced in a material that results from any changes in temperature. It is important to understand how a material will behave when heated, as stresses can cause the material to break or permanently deform.

In both metals and polymers, thermally induced stresses can be relieved by a deformation, or change in shape. Brittle ceramics, however, do not have much ability to handle thermal stress and often crack as a result of thermal shock. If you have ever put a warm glass in your freezer, you may have experienced the thermal shock of a ceramic that results from rapid cooling, as the glass will likely shatter going from warm to cold temperatures.

3.2   OPTICAL PROPERTIES

3.2.1   ORIGINS OF COLOR

You may have wondered how materials get their color. Color in materials actually arises from the interactions of light waves with individual atoms. This idea came from studying the colored flames that are emitted when metal compounds burn in what is known as a 'flame test'. For example, when lithium and strontium compounds burn, they emit a characteristic red flame, sodium burns yellow and potassium emits a lilac color.

To understand this phenomenon, we must first briefly discuss waves. A **wave** is defined as a continuously repeating change or oscillation in matter or a physical field. Light is a wave, or, oscillations in electric and magnetic fields that can travel through space. There are many forms of light, including visible light, x-rays, radio waves, and microwaves. They are all just different forms of electromagnetic radiation that vary by **wavelength** and **frequency**.

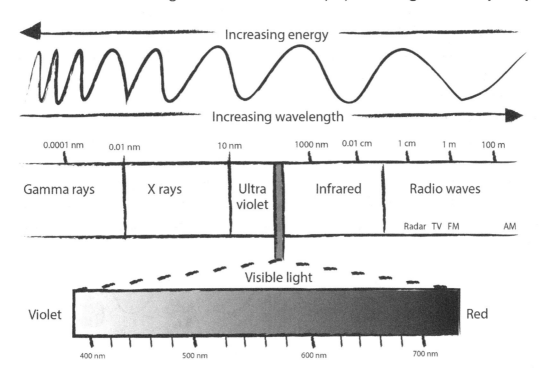

FIGURE 3.3

Figure 3.3 The electromagnetic spectrum

All waves can be characterized by their wavelength and frequency. Wavelength, often abbreviated with the Greek letter lambda (λ), is the distance between any two adjacent, identical parts on a wave. Since it is a measure of length, its unit is the meter (m).

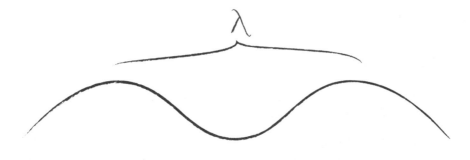

FIGURE 3.4

Figure 3.4 The wavelength distance on a wave

The frequency of a wave, abbreviated with the Greek letter nu (ν), is the number of wavelengths of a wave that pass a fixed point in one second of time. The unit for frequency is s^{-1}, or Hz. The longer the wavelength of a particular wave, the lower the frequency, and consequently, the lower the energy. Conversely, if a wave has a very small wavelength, it has a higher frequency, and thus carries more energy.

A theory known as wave-particle duality postulates that all waves can also behave as particles. As such, light can behave as both a wave and a particle. Particles of light can be thought of as tiny packets of energy that we call **photons.** Again, different wavelengths of light have different amounts of energy. The photons, correspondingly, carry more or less energy based on the size of the wave. We can summarize that the longer the wavelength of light, the lower the energy of its photons, and conversely the shorter the wavelength of light, the higher the energy of its photons.

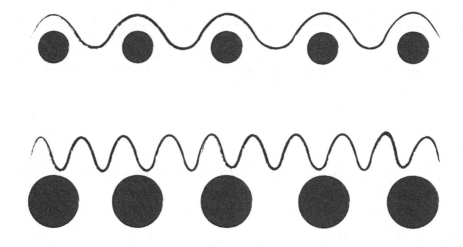

FIGURE 3.5

Figure 3.5 Light waves and their corresponding photons

3.2.2 ATOMIC TRANSITIONS

Let us now revisit the flame test. The colors from burning metals arise from electrons transitioning between different electron shells within the atom. Electrons with excess energy are called **excited electrons.** When atoms from the metals absorb heat from flames, electrons absorb some of this energy and become excited, and jump up to a higher electron shell. We know that everything in nature tends toward its lowest energy, so the electron does not remain in its excited state. Instead, the excited electron will transition back to its original lower electron shell, also known as its **ground state** energy level. According to the law of conservation of energy, energy cannot be created or destroyed. As such, the amount of energy the excited electron absorbed must be the same amount that is released upon transitioning back to the ground state. The energy released when the electron falls back to its ground state is released as a photon of light. This is where colors arise from.

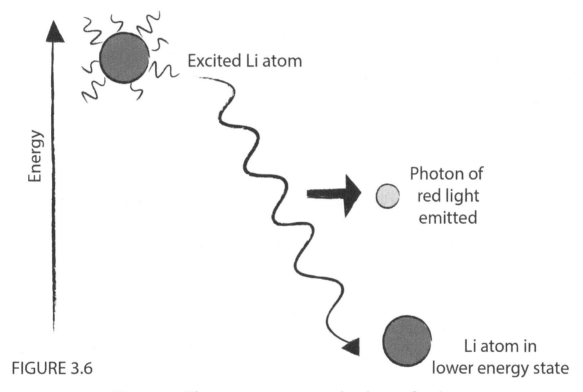

FIGURE 3.6

Figure 3.6 Electronic transition with release of a photon

3.2.3 COLORS FROM IMPURITIES

Another scenario in which colors arise from atoms absorbing light involves crystals that contain ions. In this particular case, the ions would not normally belong in the crystal, and are considered impurities.

When valence electrons are shared in covalent bonds, they become very stable, and therefore have very low ground state energy. When the ground state has such low energy,

Tania M. Cabrera

a great deal of energy is required to excite the electrons. So much energy is required, that visible light will not be able to excite these stable electrons, and in fact, it requires the higher energy of UV light to promote ground state electrons. These types of covalently bonded materials show no visible color as a result.

An exception to this class of covalently bonded materials occurs when a compound forms with transition metals. Transition metals contain incomplete outermost electron shells, so these valence electrons are easier to excite and require only visible light's energy. Compounds containing these transition metals can have very bright, vivid colors as a result.

Compounds that contain transition metal impurity ions also can have very bright coloring. One such example is a ruby. Chemically speaking, a ruby is made up of aluminum oxide (Al_2O_3). If you look at a periodic table, you will see that aluminum is not a transition metal, and will make very strong, stable covalent bonds with oxygen. It would then make sense that pure aluminum oxide, also known as corundum, is colorless. Where, then, does the dark red color come from in rubies?

In a ruby, about 1% of the aluminum atoms are replaced with chromium ions. The chromium-oxygen bond that results absorbs in the violet region at higher energy, and in the yellow-green region of the spectrum at the lower energy end. Therefore, white light, which is a combination of all colors, becomes depleted of violet and yellow-green light, leaving only red. This is exactly why a ruby appears red.

3.2.4   REFRACTION

In the previous sections, we see that color in materials can come from single atoms interacting with light. Let's now look at some examples of light interacting with bulk matter and giving rise to colors. **Bulk matter** can be defined as matter with the order of 6.02×10^{23} or more atoms. The number 6.02×10^{23} is known as a "mole" and is an important unit in chemistry.

Sometimes, color can arise from changes in the speed of light as it passes from one material to another. This change in speed of light is known as **refraction.** When light passes through a medium, the photons are absorbed by that medium, and get reemitted by any atoms that are in the light's path. This interaction slows down the light. As such, light travels fastest in a complete vacuum, or, a space devoid of matter, and more slowly in all other materials.

Let's consider a single wavelength of light passing from the air into a piece of glass. The light enters the glass from the air going at one speed, and once it enters the glass, slows down before hitting the other edge of the glass. A slower beam of light bends toward the normal, or the direction perpendicular to the edge of the glass. Once the light leaves this piece of glass, it gets pulled away from the perpendicular direction and returns to its original speed.

38

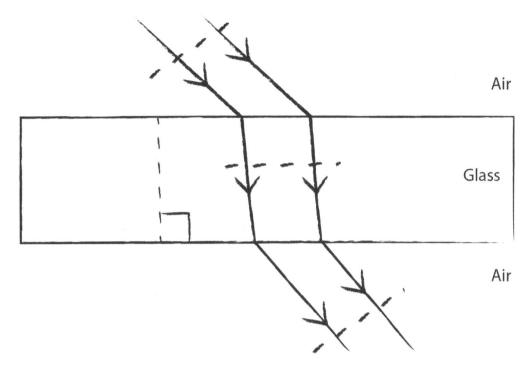

FIGURE 3.7

Figure 3.7 Light passing from air to glass and back to air

Every material has a distinct **refractive index**, or, index of refraction (n). The refractive index is a number that defines how light passes through a material obtained by Equation 3.3

Equation 3.3 $n = \frac{v}{v'}$,

where v is the speed of light in a vacuum and v' is the speed of light in that specific material.

A common example where we see the effects of refractive indices is a pencil entering a glass of water. The pencil appears to be bent or broken because the light is refracted as it enters and leaves the water.

Refraction can lead to beautiful color effects when dealing with white light, for example, the colors of a rainbow. White light is composed of a mixture of all wavelengths of colored light from red to violet. When white light is passed through a prism, refraction causes it to be dispersed, and it looks as if a rainbow is coming out of the other side. This is because different colors, or wavelengths of light have different refractive indices. The higher energy, shorter wavelengths (violet) get refracted the most, while the lower energy, longer wavelengths (red) are refracted the least.

3.2.5   INTERFERENCE

Sir Isaac Newton was the first scientist to ever document an explanation of color coming from **light interference**. Let's first explain interference patterns in light by outlining the two most extreme ways in which light waves can interact. Note that these interference interactions can apply to any kind of waves, not just light.

When two waves of the same wavelength come together to interact, the end result depends on the phase of the wave. The phase of a wave refers to where the peaks and troughs are located. If two waves of the same wavelength AND the same phase interact, the resulting wave is still of the same wavelength, but will have double the amplitude, or the sum of the two waves' amplitudes. This additive effect is known as **constructive interference**, shown in Figure 3.8.

Any two waves that are of the same wavelength but not exactly in the same phase will be out of phase. There are many ways in which two waves can be out of phase, but let's consider the example where the two waves of the same wavelength are exactly out of phase, so where one has a crest, the other has a trough. When these two waves interact, the result is a sort of "canceling out" effect, and no wave will remain. This is known as **destructive interference**, also illustrated in Figure 3.8

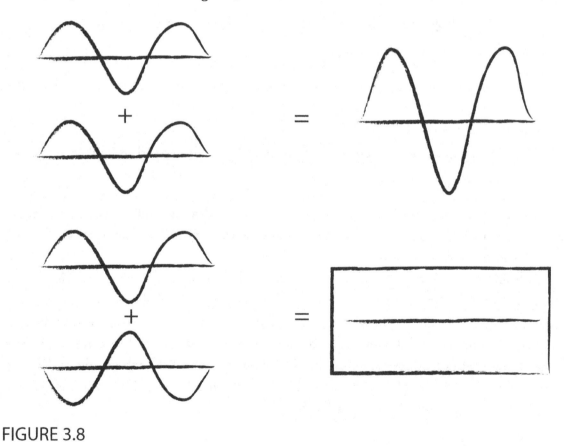

FIGURE 3.8

Figure 3.8 constructive and destructive interference

An example of interference giving rise to color effects is when light interacts with a thin film. An example of a thin film would be a soapy film, or a soapy bubble. In this case, light can be reflected from both the front and the back of the film. A bubble has variable thickness throughout. The light traveling back to our eyes from the back of the film in a thicker area will have to travel a slightly longer distance. If the light coming from the back of the film is exactly out of phase with the light coming from the front, destructive interference will be the result and no color will be seen. If the front- and back-reflected light are in phase, constructive interference will occur and color can be observed. The color and intensity of the light we see depends on the angle at which we view the thin film.

3.3 MECHANICAL PROPERTIES

Many materials are subjected to strong forces or heavy loads. It is important to understand how materials behave under such conditions to that they will not permanently fracture or deform to the point where they are no longer useful. To determine mechanical properties of materials, laboratory experiments are conducted that replicate as many conditions as possible.

3.3.1 STRESS AND STRAIN

One very common mechanical test is a **tension test**. A test material is pulled in opposite directions along one axis to see how it will deform. The results of the test are recorded as force versus material elongation. The test results will depend on the size of your test material. For example, if your test material doubles in size, it will require twice as much force to produce the same elongation. To help minimize geometric factors, both elongation and force applied are normalized to engineering strain and engineering stress. **Engineering stress** (σ) is defined by **Equation 3.4**:

Equation 3.4 $\sigma = \dfrac{F}{A_0}$

where F is the instantaneous force applied perpendicular to the test material's cross section, and A_0 is the original cross sectional area before the force F was applied.

Engineering strain (ε) is defined by **Equation 3.5**:

Equation 3.5 $\varepsilon = \dfrac{\Delta l}{l_0}$

Where $\cdot l$ is the change in length after a force is applied (final length – initial length), and l_0 is the original length before the test.

The amount a material can deform generally depends on how much force or stress is imposed on it. Hooke's Law tells us that at low levels, stress and strain are proportional to one another by the following relationship, **Equation 3.6**:

Equation 3.6 $E = \frac{\epsilon}{\sigma}$

Where E is the **modulus of elasticity**, also sometimes called the Young's Modulus. A small value for E indicates that a small stress gives rise to a large extension. These materials with small E values are quite rubbery. A material with a large Young's Modulus value is quite stiff. For example, the polymer PTFE which is used to make plastic drink bottles as an E value of 0.40-0.55 Giga Pascals (GPa), whereas a ceramic such as aluminum oxide has an E value of 393 GPa.

When stress and strain are proportional to one another, a deformation that occurs is called **elastic deformation**. Elastic deformation of a material is not permanent, meaning when the applied force to a material is released, the material returns back to its original shape. Think, for example, of pulling lightly on a rubber band and then letting go. The point at which elongation is no longer reversible is called a material's elastic limit. When stress is no longer proportional to strain, Hooke's Law ceases to be valid. Beyond the elastic limit, a permanent or, plastic deformation occurs. Imagine now pulling so hard on the rubber band that loses its shape. This would be an example of a **plastic deformation**.

3.3.2   OTHER MECHANICAL BEHAVIORS

The **tensile strength** of a material is the maximum stress a material can endure during a tension test. **Ductility** is another important mechanical property of a material. The ductility of a material refers to the degree of plastic deformation that occurs in a material at the point of fracture. A material that cannot experience much plastic deformation before it fractures would be considered very brittle.

Toughness is, in the simplest of terms, how much energy is required to fracture a material, whereas the **hardness** refers to the resistance of a material to penetration or scratching of its surface.

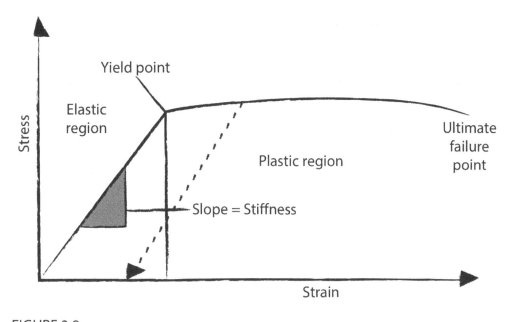

FIGURE 3.9

Figure 3.9 Graphing mechanical properties of materials

3.4   ELECTRICAL PROPERTIES

Let's now consider the electrical properties of materials, or, how materials respond to an applied electric field.

3.4.1   ELECTRICAL CONDUCTION

Electrical conduction refers to how a material transmits an electrical current. To begin to understand how materials respond to an applied electric field, we first will need to learn a little about electricity.

The most basic law pertinent to electricity is Ohm's Law, shown in **Equation 3.7** which states:

Equation 3.7 $V = IR$
where V is applied voltage (in volts), I is current (in amps), and R is the resistance of the material through which the current is passing (Ohms, ·). Often, the degree to which materials oppose electric current is referred to in terms of resistivity (ρ), or, resistance per given length:

Equation 3.8 $\rho = \frac{RA}{l}$

Where R is still the resistance, A is the cross-sectional area of the material perpendicular to the direction of the current, and l is the distance between the two points at which the voltage is measured.

If we combine resistivity with Ohm's law, we can insert R = and obtain **Equation 3.9,** which is how resistivity is experimentally measured.

Equation 3.9 $\rho = \frac{VA}{Il}$

Sometimes, we use **electrical conductivity** (σ) to describe the electrical character of a material. It is simply the reciprocal of resistivity and again tells us the ease at which a material conducts an electrical current.

Equation 3.10 $\sigma = \frac{1}{\rho}$

We can classify a material according to how easily it can conduct an electrical current. Materials will fall into one of the following categories:

- **Conductors**
- **Insulators**
- **Semiconductors**

The magnitude of a material's electrical conductivity is very strongly dependent on its chemical structure, or more specifically, how the electrons are arranged. If the material has many free electrons available to participate easily in the conduction process, like in a metal, it is considered a conductor. If the valence electrons are tied up in strong covalent bonds, as in ceramics, the material is considered an insulator. A semiconductor behaves somewhere between a conductor and an insulator.

3.4.2   ENERGY BAND STRUCTURES

As outlined in Chapter 2, we know there are a series of electron shells that electrons are allowed to occupy. These shells are often also referred to as **bands.** At an atom's ground state, the atoms have a certain energy band that contains its outermost valence electrons. This band is known as the **highest occupied molecular orbital**, or HOMO for short. There are then unoccupied energy bands above the HOMO that can be occupied by excited electrons. The next available energy band above the HOMO is called the **lowest unoccupied molecular orbital**, or LUMO for short. The ease at which valence electrons can enter these empty energy bands is what determines the electrical properties of metals.

In materials science, we use what are called **energy band diagrams** to illustrate electron energy levels, as opposed to the atom drawing seen in Chapter 2. In order for a material to be able to conduct electricity, it must be able to mobilize electrons and promote them from the HOMO to other unoccupied energy bands. In energy band diagrams, the HOMO is often called the **filled** or **valence band**. The LUMO is often called the **empty** or **conduction band.**

Since metals already contain free electrons due to metallic bonding, there is no additional energy required to make free electrons. For other materials that are insulators and semiconductors, the electrons are in covalent bonds and therefore do require energy to excite electrons from the valence band to the conduction band. To make a free electron, it has to be promoted across what is known as an energy **band gap**. This band gap is the difference in energy between the valence and conduction bands.

The larger the band gap in a material, the lower the conductivity. The only real difference between insulators and semiconductors is the size of the band gap, or how much energy is required to produce free electrons in the conduction band. If you think in terms of bonding structure, insulators are generally materials that contain strong covalent bonds. These covalently bonded electrons are in strong, stable bonds, and would require a large amount of energy to be freed. Semiconductor materials, such as silicon, do contain covalent bonds as well, but are slightly weaker. This is why with the addition of heat, semiconductors can conduct an electric current, by promoting electrons across a smaller band gap.

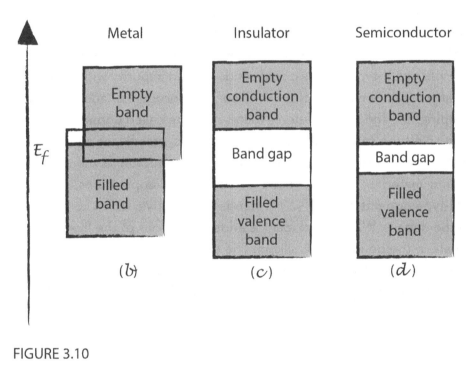

FIGURE 3.10

Figure 3.10 Energy band diagrams

3.4.3 SEMICONDUCTORS

Semiconductors are materials which have a band gap energy of less than 3 eV. There are two types of semiconducting materials that exist – intrinsic and extrinsic. An **intrinsic semiconductor** has electrical behavior inherent to the pure material. An **extrinsic semiconductor**'s electrical behavior is dictated by added impurity atoms. There are two elements on the periodic table that are intrinsic semiconductors – silicon and germanium. If you notice their location on the periodic table, you will find silicon and germanium in group four, under carbon. These elements have four valence electrons. As you move down the periodic table, atoms get larger, and therefore, the covalent bonding among atoms will get weaker. This explains why diamond, made purely of carbon, is an insulator, while silicon and germanium are semiconductors. Some other intrinsic semiconductor compounds include gallium arsenide, cadmium sulfide, and zinc telluride.

When electrons are promoted into the conduction band, what is left behind in the valence band is known as a **hole**. An electron hole is a space in a semiconductor where an electron once existed. Once an electron is promoted into the conduction band and leaves a hole in the valence band, another electron will move up to fill its place. This then leaves behind another hole, which can be filled, etc. The position of this missing electron, or hole, is therefore thought of as moving. The hole can be considered a positively charged particle that moves down as an electron moves up in an electric field.

Extrinsic semiconductors, again, are semiconductors whose behavior is determined by added impurities. Almost all commercially available semiconductors are extrinsic. There are two types of extrinsic semiconductors known as **p-type** (positive) and **n-type** (negative).

To accomplish an n-type extrinsic semiconductor, an impurity with 5 valence electrons is added to an intrinsic semiconductor like silicon. The process of adding impurity atoms is known as **doping**, and the impurity atoms themselves are called **dopants.** Possible dopants for achieving an n-type semiconductor are phosphorus, arsenic, or antimony. You may be wondering what effect these dopants have on a semiconductor. As mentioned, silicon, the semiconducting material most often used for commercial applications, contains 4 valence electrons. N-type dopants contain 5. "N" stands for negative, which comes from the extra electron in the dopant. When the dopant is added, only four of the five valence electrons can participate in the bonding with silicon. The extra, non-bonded electron can easily be promoted into the conduction band because its energy state is located within the band gap of the semiconductor material. This extra electron is called a donor.

P-type semiconductors are made by adding an impurity with only 3 valence electrons, such as boron. "P" stands for positive, and comes from the fact that the dopant has one less valence electron than the intrinsic semiconductor material. In this case, one of the covalent bonds is missing an electron, so effectively a hole is created. This impurity introduces an extra energy level within the band gap called an acceptor state.

In extrinsic semiconductors, extra electrons or holes allow for semiconductors to work at room temperature as opposed to needing heat in order to conduct electricity. As a result, these materials can be used as semiconductors for the electronic devices we use at ambient temperature, such as computers and cell phones.

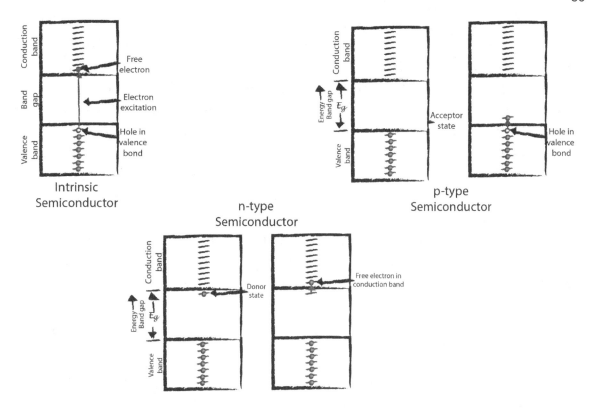

FIGURE 3.11

Figure 3.11 Energy band structures for intrinsic and doped semiconductors

3.5   MAGNETIC PROPERTIES

Magnetic behavior can be somewhat difficult to understand. There are very few materials we know of that are actually magnetic, and the ones that are magnetic are often made of iron. What makes iron so special?

If you pick up a piece of iron, it is not necessarily magnetic. If you want to magnetize this bar of iron, you may do so by placing it in a strong magnetic field. How is this magnetic field achieved? If you pass an electric current through a coil of wire, you can make a magnetic field. If you then push the iron into the coil and take it back out, you will find that it has been magnetized. Electrons going around in circles are what produce a magnetic field.

Atoms have electrons going around a nucleus, so they also produce a small magnetic field. The bigger the area the electrons are moving around, the bigger the magnetic field they produce, so it is safe to say each "atom magnet" produces an incredibly small magnetic field.

The noble gases, however, do not produce a magnetic field. This is because, as we know, they have 8 electrons in their valence electron shell. For every electron going one way, there is another going the opposite way. This causes a "canceling out" effect of their magnetic fields. It is this same canceling out effect that explains why most materials are not magnetic. If you recall, materials want filled outer electrons shells, which they achieve through bonding. If you have a full valence shell, you get canceling of magnetic fields, as in the noble gases. Why, then, do some materials manage to create a magnetic field?

If you look at the periodic table between groups IIIB and IIB, you will see what are known as the transition metals. These elements have electrons in their fourth electron shell, but their third is only partially full. When the atoms form bonds, they generally use those electrons in their fourth electron shell, leaving an unfilled third shell. This is enough to make each atom behave as a little magnet. All transition metals, however, are not magnetic.

Magnets contain a north and a south pole. In some metals, each atom points its poles in completely random directions. This causes a canceling out effect, and there is no net magnetic field. These materials are called **paramagnetic**. In other materials, the individual atom magnets arrange their poles in alternating, opposite directions. This also causes the magnetic field to cancel out. These materials are known as **antiferromagnetic**.

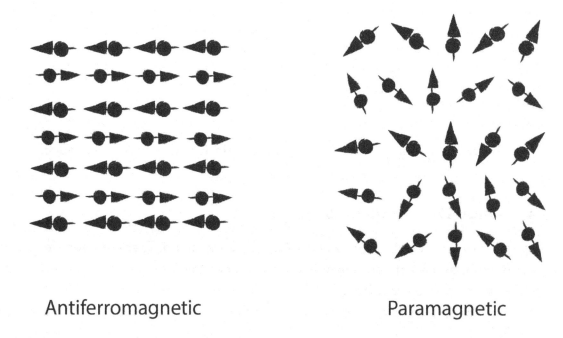

Antiferromagnetic   Paramagnetic

FIGURE 3.12

Figure 3.12 Antiferromagnetism and paramagnetism

There are three metals – iron, colbalt, and nickel – that have their atom's magnetic poles all pointing in one direction, and are called **ferromagnetic**. These elements are magnetic.

NANOSPEAK - WITH CHRISTOPHER SAVAGE (CHAPTER 3)

1. Where do you work and what is your job title?

   I am an operations principle technician at GlobalFoundries. There are approximately 800 steps in the creation of a single chip. About 10 groups monitor these steps, and the wafers run on tracks through the fab. I am involved in coordinating wafers and setting up efficient routes along the tracks. This is the best way to keep such a complicated process running smoothly.

2. What is your educational background?

   I have a bachelor's degree in business management and an associate's degree in nanotechnology.

3. How did you get interested in nanotechnology/where did you hear about it?

   I had heard about the semiconductor manufacturing industry moving into the area where I live. I really didn't like my current job in the business world, so decided to go back to school at my local community college in nanotechnology. Once I entered the program, I really enjoyed my classes and decided it was a good career path.

4. What advice would you give to someone who might want to get into the field?

   I would say that having a bachelor's degree in any engineering field will give you more job flexibility than an associate's degree. With the associate's degree, you can be somewhat limited in terms of career advancement.

CHAPTER 3 SUMMARY

- Thermal properties of a material indicate how the material responds to changes in temperature.
- The three states of matter are solid, liquid, and gas.
- Phase transitions occur when a material absorbs energy or has energy removed and changes physical state as a consequence.
- Heat capacity refers to how well a material can absorb heat from its external surroundings.
- Thermal expansion occurs upon heating materials. Depending on how the material is bonded, it may have more or less ability to expand.
- Transportation of heat from areas of high temperatures to areas of low temperature is known as thermal conductivity. This can occur via phonons or free electrons.

- Thermal stress is the mechanical stress induced in a material when it is heated or cooled rapidly. It can result in material deformation or failure.
- Color in a material may arise when atoms absorb heat from their surroundings, become excited, and relax back to their ground state with the emission of a photon.
- Color may also arise from impurities in a material; for example a ruby is red because it contains some chromium impurity.
- Refraction describes the change in the speed of light as it passes through different materials.
- When two waves interact with one another, this is known as interference. If the waves are in the same phase they may add together. This is called constructive interference. If the waves are out of phase, destructive interference may occur.
- Small particles can scatter light in different directions. Some particles may scatter different wavelengths more or less. The sky appears blue because particles in the atmosphere scatter short wavelengths (violet and blue) the most.
- Mechanical properties of materials are the result of materials experiencing applied forces.
- Engineering stress is a force applied to a material per given area. Engineering strain is the amount of displacement that occurs when a stress is applied. Stress divided by strain is a value known as Young's Modulus.
- The electrical properties of a material describe how a material responds to an applied electric field.
- Electrical conduction refers to how a material transmits an electrical current.
- Materials can be classified as electrical conductors, insulators or semiconductors. The difference between these is related to the amount of energy required to free an electron in that material.
- Naturally occurring semiconductors are called intrinsic semiconductors. Semiconductors with electrical character altered by doping are called extrinsic semiconductors.
- Very few materials are magnetic. The few that are ferromagnetic are iron, cobalt, and nickel.

REFERENCES

1. W. Callister and D. Rethwisch, *Materials Science and Engineering, An Introduction,* 9th *Ed.,* John Wiley & Sons, 2014.
2. P. Thrower, *Materials in Today's World,* 2nd *Ed.,* Learning Solutions, 1995.
3. M. White, *Properties of Materials,* Oxford University Press, 1999.
4. D. Ebbing and S. Gammon, "*General Chemistry,* 7th *Ed.,* Houghton Mifflin College Div., 2002.

CHAPTER 4 WHAT IS "NANO"

INTRODUCTION

It is important to become comfortable with the metric system and metric units for nanotechnology. It is also imperative we employ scientific notation to help us deal with the very small numbers needed for the nano scale. A number of examples will be given to help you grasp just how small the nano scale is, followed by an overview of the history of nanotechnology and nanoscale materials. This chapter will also familiarize you with some of the reasons why materials on the nanoscale are so unique and important.

Some things to think about:
- Why is scientific notation so important when dealing with nanoscale materials?
- Is a nanoparticle closest to the size of: the width of a piece of hair, a bacteria, or the diameter of a strand of DNA?
- Why have nanoscale materials received so much attention?
- Think of some new ways in which nanomaterials might be used.

4.1 UNITS

Within the scientific community, there is a need for common units used for measuring quantities such as length, temperature, and volume. While in the United States, the English system is the most widely used, the rest of the world uses the metric system.

The metric system has been the preferred system for scientists. To ensure units were kept consistent amongst the entire scientific community, an international agreement established a comprehensive system of units called **"le Système Internationale,"** or International System (commonly abbreviated SI). SI units are based on the metric system, and are summarized in the table below:

Physical Quantity	SI Unit	Abbreviation
Mass	Kilogram	kg
Length	Meter	m
Volume	Liter	L

| Temperature | Kelvin | K |

The fundamental SI units are not always of a convenient size. For this reason, the SI system has prefixes that can change the size of the fundamental unit. The most common prefixes are listed in Image 32. For example, the SI unit for length is the meter (m), but we can also use the centimeter (cm) to represent one-hundredth of a meter, or the nanometer (nm) to represent one-billionth of a meter.

Symbol	Prefix	Multiplication Factor	
E	exa	10^{18}	1,000,000,000,000,000,000
P	peta	10^{15}	1,000,000,000,000,000
T	tera	10^{12}	1,000,000,000,000
G	giga	10^{9}	1,000,000,000
M	mega	10^{6}	1,000,000
k	kilo	10^{3}	1,000
h	hecto	10^{2}	100
da	deka	10^{1}	10
d	deci	10^{-1}	0.1
c	centi	10^{-2}	0.01
m	milli	10^{-3}	0.001
u	micro	10^{-6}	0.000,001
n	nano	10^{-9}	0.000,000,001
p	pico	10^{-12}	0.000,000,000,001
f	femto	10^{-15}	0.000,000,000,000,001
a	atto	10^{-18}	0.000,000,000,000,000,001

FIGURE 4.1

Figure 4.1 Metric prefixes

4.2 SIGNIFICANT FIGURES AND SCIENTIFIC NOTATION

Because science often involves many calculations, there is a given amount of uncertainty that results when performing these arithmetic calculations. It is always important to know how much uncertainty exists in the final result of any calculation. A set of rules has been put in to place to determine how many **significant figures** the result of calculations should contain. Significant figures can be defined as all of the digits in a measured number that include all certain digits plus an additional uncertain one. For example, the ruler shown in

Figure 4.2 shows a metal rod that is between 3.8 and 3.9 cm long. A third digit should be estimated and is called an uncertain number. It is customary when measuring to record the certain numbers (3.8) and estimate one uncertain number (3.85) so we know how far past the line the measurement appears to be.

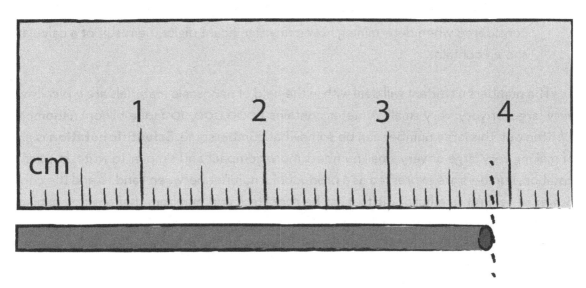

FIGURE 4.2

Figure 4.2 Measurement of a metal rod

The numbers recorded in a measurement are all significant. There are a set of rules put into place to determine which digits are significant when analyzing numbers.

1. *Non-zero integers.* Non-zeros are *always* significant. For example, the number 1,832 has four non-zero integers, all of which are significant.
2. *Zeros.* There are three different types of zeros:
    a. *Leading zeros* are placeholder zeros that come before non-zero integers. They never count as significant figures. For example, the number 0.00018 contains four leading zeros before the non-zero 1 and 8. None of the zeros in this case are significant, only the non-zero 1 and 8.
    b. *Captive zeros* are zeros that exist between non-zero integers. They always count as significant. The number 29.0056, for example, has six significant figures.
    c. *Trailing zeros* are zeros that come at the right end of a number. They are only significant if the number is written with a decimal point. The number 400 has

only one significant figure, whereas the number 400. has three significant figures.

3. *Exact Numbers.* Exact numbers are obtained by counting, for example, there are 45 cars in the parking lot, or I have 30 students in my class. It can be assumed that these numbers contain an unlimited amount of significant figures and are not considered when determining how many significant digits the result of a calculation should contain.

The numbers a student will deal with in the field of nanoscale materials are typically very, very large, or very, very small. A meter contains 1,000,000,000 (one billion) nanometers. Written out, this large number can be somewhat cumbersome. **Scientific notation** is a way of making very large or very small numbers more compact and simpler to write. In scientific notation, numbers are expressed as a product of a number between 1 and 10 and the correct power of ten. All significant digits must be reported, and numbers that are not significant should not be included when writing in standard scientific notation.

To see how scientific notation works, let's start with the example of the number 175. 175 can be written as a product

175 = 1.75 x 100
Since 100 = 10 x 10, or 10^2 so we can write
175 = 1.75 x 100 = 1.75 x 10^2

The easiest way to determine which is the correct power of 10 is to look at the original number and count the number of decimal places the decimal point needs to be moved in order to obtain a number between 1 and 10. Let's use the example 89,000,000.

The number between 1 and 10 that exists in this example is 8.9. We need to move the decimal point 7 places to the *left*, or, 7 powers of 10. This means that 89,000,000 can now be simplified in scientific notation as 8.9 x 10^7. The trailing zeros are not included as they are not significant.

Similarly, we can also represent very small numbers using scientific notation. In the case of decimals, we move the decimal point to the *right* to determine the appropriate power of 10. Look at the number 0.00045 for example. The number between 1 and 10 we can make is 4.5. The decimal place needs to be moved 4 places to the right, which requires a negative exponent. 0.00045 can be simplified as 4.5 x 10^{-4}.

4.3   HOW SMALL IS NANO?

The word "nano" comes from the Greek word "nanos," which means "dwarf." As we use the term in science today, it is a prefix of the metric system that means one billionth (1/1,000,000,000). The term **nanotechnology** refers to manmade technology or materials that contain at least one dimension whose size is on the order of one billionth of a meter, or, a **nanometer** (nm).

To gain an understanding of how small one nanometer is, consider that the width of a human hair is 100,000 nm across, the same as about the thickness of a sheet of paper. Here are some other examples:

There are 25,400,000 nm in one inch.
A strand of human DNA is about 2.5 nm in width.

In the second it takes a man to pick up a razor and bring it to his face, the facial hair will grow approximately one nanometer.

To make a comparison of scale, if the diameter of a marble were one nanometer, then the diameter of the entire earth would be a billion times larger, or, one meter.

4.4   HISTORY OF NANOTECHNOLOGY

Nanotechnology has actually existed for thousands of years. As will be explained in the following section, nanoscale materials have very different properties from their bulk counterparts, for example, gold and silver nanoparticles are not actually gold and silver colored. The Romans were combining gold and silver nanoparticles into glass to achieve beautiful colors as many as two thousand years ago. While we are not sure how they made bulk metals into nanoparticles, we do know that they were able to create glass ranging in colors from red to yellow to blue. Perhaps the most famous example of this Roman glass nanotechnology is the Lycurgus Cup, which now resides in the British Museum. The Lycurgus Cup contains gold and silver nanoparticles, and can appear red or green, depending on which direction light passes through it.

As far back as 300 BC, Arab craftsmen also incorporated nanotechnology into steel when creating their famed Damascus steel swords. These Arab swords were of legendary strength, and known for distinctive water-like patterns. A German research team in 2006 confirmed that the steel contained both carbon nanotubes and nanowires accounting for its strength. The synthetic methods are still unknown.

The Renaissance brought about continued use of nanomaterials, although the presence of materials with sizes of less than 100 nm was still not known or characterized. Italian pottery from the 16th century with vivid colors has been confirmed with modern micrographs to contain metallic nanoparticles. It would not be until the 1800s when British physicist and chemist Michael Faraday learned to use gold to synthesize gold nanoparticles. While the tools to see and characterize such materials still did not exist, Faraday noticed that solutions of these gold nanoparticles could have different colors depending on their very small size.

The very famous German physicist Albert Einstein later explained in 1905 that these gold particles would never settle out of their solutions because materials on the nanoscale have negligible mass and therefore do not follow the rules of gravity as do bulk materials.

Gold nanoparticles would move around in solution forever before being impacted by solvent molecules in what is described as "**Brownian Motion**."

In the years that followed Einstein's explanation of gold nanoparticles in solution, Austrian-Hungarian chemist Richard Adolf Zsigmondy won a Nobel Prize for his continued work with colloidal gold, and was the first scientist to use the term *nanometer* to describe these small particles. Additionally, the German physicist Ernst Ruska developed the first transmission electron microscope, which allowed visualization at the nanoscale for the first time in 1931.

The beginning of our any ability to control materials at the nanoscale was first demonstrated in 1932 by Brooklyn-born chemist and physicist Irving Langmuir and Schenectady, New York-born physicist Katherine Blodgett. Langmuir and Blodgett worked at General Electric in Schenectady, New York and together discovered a method by which a single molecular layer could be deposited onto a substrate. This single layer, known as a monolayer, was given the name **Langmuir-Blodgett film**. Langmuir-Blodgett films can be made by floating an organic, or carbon-containing molecule, on the surface of a liquid and immersing a solid substrate into the liquid. As the solid substrate is pulled up through the floating organic molecules, they will adsorb onto its surface one layer at a time.

In December of 1959, American physicist Richard Feynman gave a very famous talk at Caltech called "There is Plenty of Room at the Bottom." In this famed talk, Feynman said that the ability to manipulate matter at the atomic, or nanoscale, would become the most powerful form of synthesis to date and create many new technological opportunities. He asked such intriguing questions as, "how small can you make machinery," "how small can we make writing," and "what is the ultimate limit?" Feynman challenged scientists to construct a tiny motor, and to fit the entire contents of the Encyclopedia Brittanica on the head of a pin, which would require a 20,000x reduction in linear size. Both challenges were met in 1960 and 1985 respectively.

After Feynman's 1959 talk, the term "nanotechnology" was coined by Japanese professor Norio Taniguchi, and a new era in which nanotechnology was well understood and rapidly expanding began. Many advancements were made, including the discovery of carbon 60, or the "buckyball" in 1985 by Harold Kroto, Robert Curl, and Richard Smalley, as well as the discovery of carbon nanotubes in 1991 by Sumio Iijima.

Other advancements included the first semiconductor nanoparticle quantum dots solutions by Louis Brus, man-made molecular motors, fuel cell fabrication using nanoscale materials, and many commercial applications of nanotechnology.

4.5 HOW ARE NANOSCALE MATERIALS SO UNIQUE?

Now that we have a feel for the history and development of nanoscale technology, we can begin to learn what make nanoscale materials so different and unique from their bulk counterparts. In other words, what makes these tiny materials so special and receive so much attention?

Perhaps the most obvious answer to our question about the uniqueness of nanoscale materials is the knowledge that nanoscale materials are simply incredibly small. The smaller a material is, the more we can fit into a small space. The most profound example of this is in the field of microelectronics and silicon computing chips. Electronics work by a series of opening and closing switches called **transistors**. The more transistors we can fit on a chip, the faster and more powerful and affordable our electronic devices can become, as well as more affordable. In the 1970s, a microchip would contain about 2,000 transistors. With the advent of nanotechnology, chips are approaching the 2 billion transistor mark. At the transistor per chip cost of the 1970s, an Apple iPod® would cost around 3.2 billion dollars.

Another feature of nanoscale materials that gives rise to their unique properties is the enormous **surface area to volume ratio**. The **surface area** of a material will reduce with size. Imagine for example, a cube of metal that is 2 x 2 x 2 cm. You could calculate the surface area with the formula

Surface area = $6a^2$
Where a = 2.
The resulting surface area would be
Surface area = $6 \times (2)^2 = 24$ cm^3

Now imagine cutting that cube down the middle. The result would be two rectangular prisms, each 2 x 2 x 1 cm. The formula for the surface area of each rectangle is

Surface area = $2ab + 2bc + 2ac$
The resulting surface area for each prism would ne
$2(2)(2) + 2(2)(1) + 2(2)(1) = 8 + 4 + 4 = 16$ cm^3

Adding the surface area of *both* rectangular prisms together would give

16 cm^3 + 16 cm^3 = 32 cm^3

As you can see, the smaller we make materials, the more surface area becomes exposed. Nanoscale materials, as a result, have a very high surface to volume ratio.

Surface Area to Volume Ratio = $4{\cdot}R^2 / \frac{4}{3}{\cdot}R^3 = 3/R$
where R is the particle radius.

The smaller the value for R, the larger the surface area to volume ratio of the material. This is significant because as a particle gets smaller, the larger the percentage of its atoms exist as surface atoms. Atoms at the surface of a material influence its chemical and physical interactions with the environment, like in chemical reactivity. For example, while bulk grain is not at all flammable, grain dust can be explosive. On the nanoscale, we can compare bulk gold with gold nanoparticles. Picture bulk gold, widely used as jewelry because it is inert and will not oxidize when exposed to air as other metals do. It is shiny, gold in color, and can conduct electricity well. Nanoparticles of gold, on the other hand, are actually never gold in color. Depending on the diameter of the nanoparticles, they can range in color from red to green to orange. Interestingly, gold nanoparticles are not inert like their bulk counterpart, but are in fact quite reactive and often used as catalysts to lower the energy needed to perform various chemical reactions. They also do not conduct like a metal would, and are instead semiconductors.

4.6   HOW ARE NANOMATERIALS BEING USED?

As outlined in the previous section, materials on the nanoscale are very unique and interesting and give rise to many opportunities for novel applications. Nanotechnology is very interdisciplinary and has roots in math, physics, chemistry, biology, and engineering, all of which draw from nanotechnology for exciting new achievements.

Biologists have employed nanotechnology, as many biological structures are also on the nanoscale. Integrating nanotechnology with biology has led to many new developments, including disease diagnostics, magnetic resonance imaging (MRI) contrast agents, and targeted drug delivery mechanisms, to name a few. Nanotechnology has even been used to repair damaged human tissues by growing them on scaffolding made of nanoscale materials.

There is the potential to apply nanotechnology to improve our environment. For nanomaterials can be used to purify contaminated ground water and clean our air. We can also utilize nanotechnology in lowering our energy consumption with efficient light emitting diode (LED) devices, new batteries and fuel cells. New lightweight solar panels can be manufactured that are printed on flexible material that require less energy to synthesize than traditional heavier panels.

We also see an increase in the use of nanotechnology in everyday items such as automobiles and sporting goods. Different car manufacturers are leaning toward vehicle materials that are strong, lightweight, and very efficient. Sporting goods companies that make items such as golf clubs, tennis racquets, and bicycles have all integrated nanotechnology into their designs. There are new stain-resistant paints and fabrics that incorporate nanoscale materials and can be found for consumer purchase.

4.7   THE FUTURE OF NANOTECHNOLOGY

The truth is that nanotechnology is still in its formative phase, but is maturing very rapidly. Between 1997 and 2005, investment in nanotechnology research and development by governments around the world increased from $432 million to about $4.1 billion. It is estimated that by 2018, products incorporating nanoscale materials will have contributed approximately $3.3 trillion to the global economy. About two million workers will be employed in nanotechnology-based industries, and perhaps three times that many will have supporting jobs.

NANOSPEAK – WITH ANTHONY LEONETTI (CHAPTER 4)

1. Where do you work and what is your job title?
    I work for Applied Materials, and I am a Customer Engineer.

2. What is your educational background?
    I have a high school diploma and have nearly completed my associate's degree in nanoscale materials technology.

3. How did you get interested in nanotechnology/where did you hear about it?
    I read a lot of technology blogs online and became very interested in nanotechnology through reading.

4. What advice would you give to someone who might want to get into the field?
    I would suggest trying to get an internship so you get a feel for what you like. For example, if you like more of the mechanical or technical side of semiconductor manufacturing, you would want to get more of a mechanical engineering background. If you became more interested in process development, a degree in physics, chemistry, or nanotechnology may be more helpful.

CHAPTER 4 SUMMARY

- The metric system is the preferred system for scientists.
- SI units are used to keep measurements consistent. The SI unit for length is the meter. We can use various prefixes to indicate smaller or larger divisions of a meter.
- Significant figures are important for ruling out uncertainty in measurement.
- The word "nano" comes from the Greek word for dwarf, and as a prefix means one billionth, or 1×10^{-9}.
- The history of nanotechnology predates when we first began to understand these tiny materials.
- Nanomaterials are special because you can fit so many into one area and because their physical properties are different from bulk materials.
- Nanoscale materials are being used for any number of applications in different fields such as medicine, alternative energies, and every day consumer items.

REFERENCES

1. S. Zumdahl and D. Decoste, *Introductory Chemistry: A Foundation, 7th Ed.,* Brooks Cole, 2010.

2. A. Einstein, *Investigations on the Theory of Brownian Movement,* Dover Publications, 1956.

3. R. Zsigmondy, "Ueber wässrige Lösungen metallischen Goldes," *Justus Lieblings Annel der Chemie,* **301 (1),** 1898, p. 29

4. R. Feynman, *Plenty of Room at the Bottom,* American Physical Society, 1959.

5. H. Kroto, J. Heath, S. O'Brien, R. Curl, R. Smalley, "C_{60}: Buckminsterfullerene," *Nature,* **318,** 1985, p.162.

CHAPTER 5   CARBON NANOMATERIALS

KEY TERMS

QUANTUM MECHANICAL MODEL
ORBITALS
IONIC BOND
COVALENT BOND
VALENCE BOND THEORY
HYBRID ORBITALS
ALLOTROPE
AMORPHOUS

DELOCALIZED ELECTRONS
BUCKMINSTERFULLERENE
ANTIOXIDANT
FREE RADICAL
FUNCTIONALIZATION
CARBON NANOTUBE
GRAPHENE

INTRODUCTION

Carbon-containing materials are ubiquitous on our planet. Of the more than 50 million compounds presently know, the majority of them contain carbon. In this chapter you will learn about carbon's special electronic. This electron arrangement gives carbon the ability to make four covalent bonds, and in turn an immense number of possible combinations for forming compounds.

You will learn about some of the bulk forms of pure carbon materials, and then will read some history as to the discovery of the nanoscale carbon materials. You will then learn about the unique properties of these carbon nanomaterials, as well as how they can be used in everyday applications.

Some things to think about:
- How is the quantum mechanical model of the atom different from the electron shell model?
- Why do carbon atoms participate in covalent bonding?
- Why is the electrical character of diamond different from graphite?
- What are some of the special properties of carbon nanomaterials?

5.1   ATOMIC STRUCTURE IN TERMS OF ORBITALS

A simplified version of atomic structure was outlined in Chapter 2. To better understand carbon chemistry, let's take a look at the actual way in which electrons are organized in an atom. While the model given in Chapter 2 helps to explain bonding, the correct model is slightly more complex, and is known as the **quantum mechanical model**. According to the quantum mechanical model of the atom, the exact location of electrons can never be determined. Instead, their behavior is described in terms of a mathematical equation known as a wave equation. These wave equations give rise to what are known as wave functions, or

orbitals. Orbitals are mathematical probabilities that describe a volume of space around an atom's nucleus that an electron is most likely to occupy. While these orbitals have no well-defined boundaries, we can say that an orbital represents the space an electron occupies about 95% of the time.

There are four different kinds of orbitals, each having a different shape. The four oribitals are designated *s*, *p*, *d*, and *f*. For our understanding of carbon materials, we will mostly concern ourselves with the *s* and *p* orbitals.

An *s* orbital is in the shape of a sphere, with the atom's nucleus at its center. A *p* orbital is dumbbell-shaped. There are 5 *d*-orbitals, 4 of which are look like a 4-leafed clover, and one which looks like a dumbbell surrounded by a donut. There are 7 *f* orbitals which have very complex shapes.

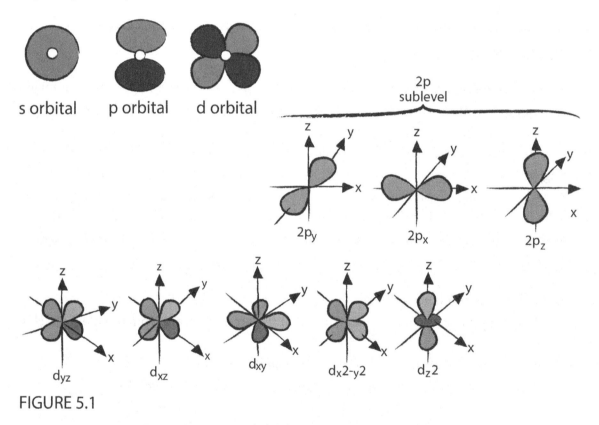

FIGURE 5.1

Figure 5.1 *s*, *p*, and *d* orbitals and their orientations

5.2   CHEMICAL BONDING THEORY

Chapter 2 explains that atoms bond together in order to attain full valence shells like those of noble gases, which are more stable and lower energy configurations than partially filled shells. The group one elements, for example, can attain this configuration by losing one electron, and the group 7 elements can attain noble gas configuration by gaining one electron. The charged atoms, or ions that result, are held together in an attraction we describe as an **ionic bond**.

Elements closer to the middle of the periodic table, however, do not participate in ionic bonding, as it would require too much energy for them to gain or lose too many electrons to look like noble gases. These elements participate in sharing of electrons to form **covalent bonds**. Again, the number of covalent bonds formed will depend on how many electrons are required to achieve noble gas configuration. Since carbon has four valence electrons, it needs four more to look like neon, so will form four bonds.

How, exactly, does this sharing of electrons lead to a chemical bond? One theory that describes covalent bonding is known as the **valence bond theory**. According to this theory, a covalent bond forms when two atoms get very close to one another and an occupied orbital overlaps another occupied orbital of a different atom. The electrons get paired in these overlapping orbitals and are attracted by the positively charged nuclei of both atoms. The simplest example of this is in a molecule of hydrogen, or H_2. In the example of H_2, the hydrogen-to-hydrogen covalent bond results from the overlap of two hydrogen s orbitals.

The covalent bonding that takes place between carbon atoms is slightly more complicated. As we have mentioned, carbon has four valence electrons and therefore makes four covalent bonds. Carbon uses two kinds of orbitals – s and p – when bonding, and as such, one might imagine a carbon bonded to four others might have two different kinds of bonds. Research shows, however, that all four bonds are exactly the same. How can this be explained?

American chemist Linus Pauling showed in 1931 that s and p orbitals from one atom can combine, or hybridize, to form four atomic orbitals that are the same. These **hybrid orbitals** are oriented in a tetrahedron and are called sp^3 hybrids. The term sp^3 tells us that the s orbital has combined with 3 p orbitals, not how many electrons occupy the orbital.

Linus Pauling's idea about hybridization gives us an explanation of how carbon can form four bonds that are the same. The shape of the hybrid orbitals can tell us why. When an s orbital combined with the three p orbitals, the sp^3 hybrid orbitals are not symmetrical. Instead, one lobe is larger than the other. This larger lobe can effectively overlap with an orbital from another atom to form a covalent bond. This more effective overlapping explains why hybrid orbitals make stronger bonds than do unhybridized orbitals.

Each of the four hybrid sp^3 orbitals from one carbon atom can overlap with other sp^3 hybrid orbitals in other carbon atoms to form 4 identical bonds. Figure 5.3 illustrates this overlapping of hybrid orbitals to form an sp^3-sp^3 carbon-carbon bond. To make the image clearer, only the larger lobes are shown.

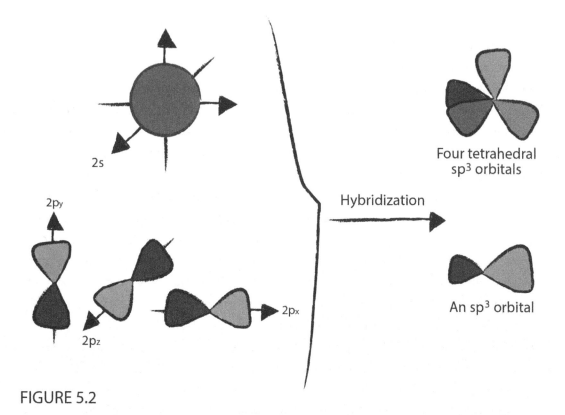

FIGURE 5.2

Figure 5.2 Four sp^3 orbitals and sp^3-sp^3 carbon-carbon bond

In a second type of carbon-carbon bond, the 2s orbital from a carbon atom combines with only two of the three p orbitals. This will result in three sp^2 hybrid orbitals and one unhybridized p orbital. Just like sp^3 orbitals, sp^2 hybrid orbitals are not symmetrical, with a larger lobe that can overlap with others to make strong covalent bonds. The three sp^2 hybridized orbitals lie in a plane 120° from one another with the remaining unhybridized p orbital perpendicular to the sp^2 plane.

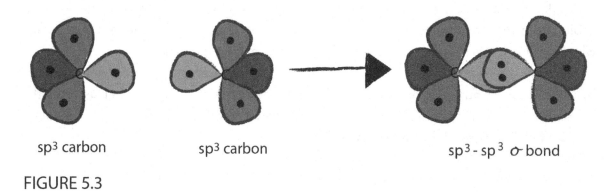

FIGURE 5.3

Figure 5.3 sp^2 hybrid orbitals

5.3 BULK CARBON MATERIALS

There are several different **alloptropes** of bulk carbon that exist. The term allotrope refers to an element that can exist in different forms. In other words, an allotrope is a material made of the same element only bonded differently.

The best known bulk allotropes of carbon are diamond, graphite, and amorphous carbon. All three allotropes are chemically and physically very unique. Diamond contains sp^3 hybridized carbon atoms covalently bonded to four others in a tetrahedral formation. Physically, diamond is transparent, colorless, and the hardest naturally occurring material on earth. While it can conduct heat via lattice vibrations, it is an electrical insulator since all valence electrons are participating in sp^3-sp^3 carbon-carbon bonds. Diamonds are rare and very expensive, commonly used in jewelry and for cutting applications.

Another allotrope of carbon we commonly see is graphite. Graphite is the material used in pencils, commonly called "pencil lead." Pencil lead is somewhat of a misnomer, since it actually contains no lead atoms at all, only carbon. Graphite is opaque, black, and fairly soft. It can conduct both heat and electricity relatively well, and is very inexpensive.

How can two materials made up of the same type of atoms be so different? The answer lies in the bonding structure of graphite. As mentioned, diamonds contain carbon atoms covalently bonded to 4 others in a three-dimensional tetrahedral formation. Graphite, on the other hand, has sp^2 hybridized carbon bonded to three others. We said that in sp^2 hybridization, the s orbital combined with 2 of the p orbitals, leaving one p orbital unhybridized. The carbon atoms make 2 dimensional sheets with hexagonal arrangement of the carbon atoms, similar to chicken wire. In between these planar layers of carbon, there are no covalent bonds, only the electrons in the unhybridized p orbital free to move between atoms. These unhybridized electrons are known in chemistry as delocalized electrons, and are the reason sp^2 hybridized carbon forms such as graphite are able to conduct electricity. The lack of covalent bonding between the sheets also accounts for the softness of graphite, as the two-dimensional planes can easily slip over one another.

The final bulk allotrope of carbon is known as amorphous carbon, or soot. Soot is relatively impure, and has no regular crystal structure. The term **amorphous** refers to a material with no long range crystalline order, or no regular arrangement of its atoms. Amorphous carbon contains a mixture of both sp^3 and sp^2 hybridized carbon atoms. While we may not often think of soot as being useful, it does have many applications, often as a colorant for items such as ink and cosmetic mascara.

5.4 BUCKYBALLS

For thousands of years, it was believed that diamond and graphite were the only pure allotropes of carbon that existed on our planet. Scientists thought they knew everything there was to know about carbon, how it bonded, and formed compounds both with itself and with other atoms.

The idea that there could be new pure carbon allotropes began by looking to outer space. Dying stars were found to emit carbon atoms, and it is this star dust that is the source of all carbon in the universe, including the carbon atoms that make up human life. Since we are all made of carbon stardust, researchers began to wonder what its structure was.

All molecules absorb light, and when this light absorption is studied in the laboratory, each molecule produces a unique absorption spectrum. Scientists in the 1980s, began to study stardust, and were surprised by the molecular absorption spectrum they obtained.

In 1982, the American physicist Donald Huffman from the University of Arizona traveled to the Max Plank Institute for Nuclear Physics in Heidlberg, Germany. There, he began to study the ultraviolet light absorption spectrum of carbon dust with German physicist Wolfgang Kratschmer. Their experiments involved pushing electric current through sticks of graphite inside of a bell jar. This produced clouds of carbon dust, similar to the stardust produced in space. When the ultraviolet absorption spectrum of this carbon dust was studied, a consistent yet unexpected band at 220 nm was continually seen. The band had a double hump and as such was dubbed the "camel spectrum." Huffman and Kratschmer were unsure if this band was something new or just "junk." They hypothesized it could be long chains of carbon atoms.

Meanwhile, British chemist Harry Kroto was also studying stardust and searching for long chains of carbon atoms at the University of Sussex. In 1984, Kroto traveled to Rice University in Texas to team up with American physicist Richard Smalley. Smalley had an instrument that would vaporize materials such as silicon. The vaporized materials would arrange into clusters and were then sent to a mass spectrometer where the mass of these clusters could be analyzed by counting the number of atoms present in each cluster. While Smalley was primarily working with silicon, Kroto used the instrument to vaporize graphite and simulate conditions in the stars.

Around the same time, fellow American chemist at Rice University Robert Curl also saw evidence of long chain carbon molecules as in the experimental results of Kroto and Smalley, but consistently saw clusters of 60 carbon atoms. Experiment after experiment, it seemed 60 was the cluster that carbon preferred. The question to be answered was: what was so special about 60?

We know that carbon likes to bond with 4 other carbon atoms, as the sp^3 hybridized carbon atoms in diamond do. At the ends of a diamond lattice, however, the carbon atoms will bond to hydrogen atoms. If you are wearing a diamond ring on your hand, for example, and run your finger across the surface, what you are touching is actually a layer of hydrogen, not carbon. The same is true of sheets of graphite, where carbon atoms are sp^2 hybridized and bond to three other carbon atoms. At the ends of the carbon lattice, the carbon atoms will bond to hydrogen. The results of Curl, Kroto, and Smalley's experimental peaks at 60 showed absolutely no hydrogen - only carbon. This made them hypothesize that the cluster at 60 must have a closed cage structure.

The question that now needed answering was how could carbon form a closed cage structure? Kroto thought of American architect Buckminster Fuller. Fuller was famous for making lightweight spherical buildings known as geodesic domes. Kroto had been inside one of Buckminster Fuller's geodesic domes at the 1967 International and Universal Exposition (Expo 67) in Montreal, Canada. He remembered that the dome was primarily made up of hexagons, but it also contained pentagons. The hexagonal shape will curve into a sphere only when placed around pentagons.

Using paper hexagons and pentagons, the shape of C_{60} emerged, with 12 pentagons and 20 hexagons of sp^2 hybridized carbon atoms. The shape of C_{60} is comparable to that of a soccer ball. They decided to name the structure **Buckminster Fullerenes** or "Buckyballs" for the geodesic dome architecture of Buckminster Fuller. C_{60} was a perfectly symmetrical cage of carbon 1 nm wide.

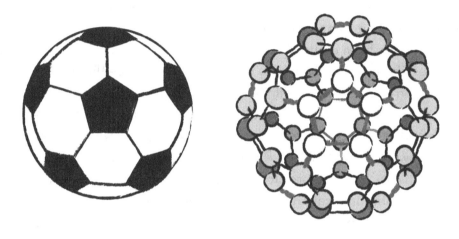

FIGURE 5.4

Figure 5.4 C_{60}

In 1996, Curl, Kroto, and Smalley were awarded the Nobel Prize for their discovery of Buckyballs. This discovery was of incredible significance since it opened the door to the idea that new pure forms of carbon can and do exist, and Buckyballs were one of the first nanoscale materials discovered. Carbon clusters of all sizes were subsequently named Buckminsterfullerenes, or fullerenes for short. Kratschmer and Huffman later discovered a method in which large, gram-sized quantities of Buckyballs could be synthesized.

C_{60} contains sp^2 hybridized carbon atoms and are sized on the nanoscale. As a result, Buckyballs have very unique properties. The delocalized electrons from unhybridized p orbitals allow Buckyballs to act as an **antioxidant**. An antioxidant is a molecule that can stop the oxidation of other molecules that result from the production of **free radicals**. Free radicals are atoms or molecules with unpaired electrons. Since electrons like to exist in pairs, they are incredibly reactive and start a chain reaction forming more and more new free radicals. When this chain reaction occurs in our cells, it can cause cell death and aging, or damage in the form of DNA mutation which can lead to cancer. An antioxidant can end free radical chain reactions by removing free radicals, or accepting unpaired electrons. Buckyballs are good electron acceptors, and when they encounter free radicals, the unpaired electron joins with a delocalized electron in a Buckyball.

While C_{60} is not inherently water-soluble, its surface can be chemically altered in a process known as **functionalization**. One example of changing the surface of C_{60} is called the Bingel reaction, named for German chemist C. Bingel in 1993. The Bingel reaction allows molecules to be added to C_{60}'s surface that can easily be altered to change the solubility of Buckyballs for biological applications.

In addition to acting as effective antioxidants, buckyballs are also incredibly strong. They have a hardness similar to or greater than that of diamonds. There has been interest in using Buckyballs to strengthen plastics or polymers, develop strong new coatings, and even plans to use them in body armor.

When elements from group 1A or group 2A on the periodic table are placed between C_{60} molecules, there is a total absence of electrical resistance that occurs. This means that an electrical current can flow indefinitely with no loss of power. If this technology were to be used in our power grid, there would be the potential to eliminate power loss and save money. Currently, however, the technology is somewhat primitive, expensive, and only works at incredibly lowered temperatures. Research on how to get these types of Buckyball wires to conduct at higher temperatures to efficiently carry current is ongoing.

Another potential application of C_{60} is in the field of medical imaging. When undergoing an MRI, a contrast agent is often used to improve the visibility of the scan. The current contrast agents used are made of the element gadolinium. While gadolinium compounds are generally considered safe for most patients, there is the possible risk of rare, but serious side effects that involve formation of excess tissue in the skin, joints, eyes, and other internal organs. It is thought that if the gadolinium was caged inside of a Buckyball, this could improve both the effectiveness and safety of current contrast agents.

SECTION 5.5: CARBON NANOTUBES

The history of the **carbon nanotube** is still not completely clear, and credit to their discovery can be a topic of debate amongst scientists. After discovering the structure of buckyballs, Richard Smalley realized in 1990 that in theory, a tube-shaped fullerene could be possible with only sp^2 hybridized carbon hexagons. American physicist and Massachusetts Institute of Technnology professor Mildred Dresselhaus dubbed these fullerenes "Buckytubes."

It seems, however, that carbon nanotubes may have been discovered some thirty years earlier without being recognized or appreciated at the time. In the 1950s, an American physicist working at Union Carbide, Roger Bacon, was studying carbon under various temperature and pressure conditions, specifically near its triple point. He observed hollow tubes of carbon that appeared to be like layers of graphite rolled upon themselves. Bacon published his findings, and did show the carbon tube structure in one of his paper's figures, but is not currently credited with the discovery of carbon nanotubes.

In the 1970's, Japanese chemist Morinobu Endo once again saw these carbon tubes, but it would not be until 1991 after the structure of fullerenes had been confirmed that Japanese physicist Sumio Iijima observed pure carbon nanotubes.

As in Buckyballs, carbon nanotubes contain sp^2 hybridized carbon atoms covalently bonded to three others in a hexagonal shape. Their diameter is also 1 nm, but the tubes can be quite long, even on the centimeter scale. This length-to-diameter ratio of up to $1.32 \times 10^8 : 1$ is higher than any other material on earth. Different types of Buckytubes exist. Some carbon nanotubes are cylindrical with open ends, similar to a straw. Others are "capped" on the ends with a half of a Buckyball.

Some carbon nanotubes exist as multiple tubes within tubes, and these are known as multi-walled carbon nanotubes, while others are one individual tube known as a single-walled nanotube. Carbon nanotubes have some unique properties, and perhaps most interesting are their thermal, mechanical, and electrical properties.

Both diamond and graphite conduct heat well via lattice vibrations, or phonons. Nanotubes are also very thermally conductive when heat is applied lengthwise along the tube. Measurements have shown that carbon nanotubes have a room temperature thermal conductivity of around 3000 W/m K. Compare this value to the well-known quality thermal conductor copper, whose thermal conductivity has been measured at 385 W/m K. In the other direction, perpendicular to its vertical axis, a carbon nanotube's thermal conductivity is about 1.5 W/m K, similar to that of insulators such as granite or limestone.

The electrical conductivity of carbon nanotubes will depend on the way in which the carbon atoms are oriented in the tube. There are three different orientations that a carbon nanotube can take: Zigzag, chiral, and armchair.

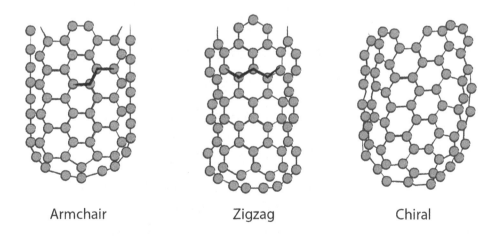

Armchair   Zigzag   Chiral

FIGURE 5.5

Figure 5.5 Three orientations of carbon nanotubes

If the carbon atoms of a nanotube are in armchair orientation, the nanotube will behave as a good electrical conductor like a metal. To make a comparison with copper again, armchair carbon nanotubes can carry an electrical current density of 4×10^9 A/cm², while copper's maximum current density is about 1,000 times less at 3.1×10^6 A/cm². Nanotubes in zigzag and chiral orientation are semiconducting. Recall from Chapter 3 that in semiconductors, energy must be added to the material in order for it to conduct an electrical current.

Carbon nanotubes also have very interesting mechanical properties. There are no other materials we have discovered that have a tensile strength, or Young's (elastic) modulus, as high as carbon nanotubes. As such, they are some of the strongest materials on earth. They have an incredibly low density since they are hollow, and are also bendable.

You can imagine that there are many potential applications for such an interesting material. Carbon nanotubes have been used to add strength and lightness to materials such as sporting goods like bicycles and golf clubs, as well as vehicles such as cars and ships. Carbon nanotubes have also been mixed with epoxy to make composite materials that are around 30% stronger than any other composites known. There have been transistors made with individual carbon nanotubes, and nanotubes made into electrical cables and wires. In the field of alternative energy, carbon nanotubes have been used in flexible batteries, solar cells, ultracapactors and for storing hydrogen.

5.6 GRAPHENE

Graphene is another pure carbon allotrope that is a single, planar sheet of graphite. As a single sheet of graphite, graphene contains sp^2 hybridized carbons covalently bonded in a two-dimensional hexagon pattern.

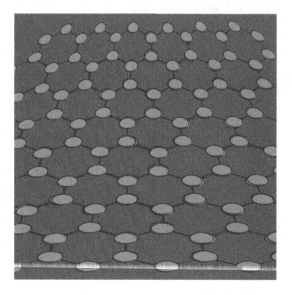

FIGURE 5.6

Figure 5.6 Graphene

The term graphene comes from a combination of the word *graphite* with the chemistry suffix *-ene* by German chemist Hanns-Peter Boehm who described single carbon layers back in the 1960s. Russian-born physicists Andre Geim and Konstantin Novoselov were awarded the 2010 Nobel Prize for discovering a simple method for isolating these two-dimensional carbon sheets in 2004.

Graphene has a very high tensile strength and can add strength when combined with other materials. While it is atomically thin and would require more than a million graphene sheets stacked upon one another to reach a millimeter of thickness, if a single sheet of graphene were stretched over a ceramic coffee mug it would be strong enough to support the weight of a car.

Graphene also has unique electrical properties. It is a conductor with the highest electron mobility of any material which is the speed at which electrons travel through the material at room temperature. There is the potential to create graphene transistors that would be much faster and smaller than those currently built using silicon. There is a physical limit to how small we can make features with silicon, and so it is estimated that by the year 2020 or 2025, this technology will need to be replaced.

Graphene shows great promise, however there are some issues when making graphene devices. Once problem is that graphene's properties are easily lost when the sheets are bent or put down onto another material. Researchers will need to figure out what is the best way to handle and manipulate such an extremely tiny material. There have been some studies that show you can cut graphene using hot nanoparticles of silver. If the graphene sheet is cut into strips known as ribbons, the ribbons are semiconducting. As such, there is great potential for the invention of new devices that currently do not exist.

NANOSPEAK – WITH JAMES CUERDON (CHAPTER 5)

1. Where do you work and what is your job title?
   I work as a process technician in chemical-mechanical polishing (CMP) at The College of Nanoscale Science and Engineering.

2. What is your educational background?
   I have a high school diploma and an associate's degree in nanoscale materials.

3. How did you get interested in nanotechnology/where did you hear about it?
   I was enrolled in the storage battery technology program at my local community college. One of the required courses was materials science. While taking this course I learned about the many job opportunities in nanotechnology, so I switched majors.

4. What advice would you give to someone who might want to get into the field?

Anyone interested in math and science *should* enter the field of nanotechnology! It is still in its infancy and is only going to expand and get bigger. There is a lot of room to grow in a career in this field.

CHAPTER 5 SUMMARY:

- The correct model for atomic structure is the quantum mechanical model wherein the exact location of electrons cannot be determined, but a probability of their location can be calculated.
- The mathematical probability of where an electron is likely to reside is called an orbital.
- There are 4 different types of electron orbitals: s, p, d, and f.
- Carbon is in group 4A of the periodic table, so it contains 4 valence electrons and can make 4 covalent bonds.
- The predominant theory for how covalent bonding occurs is the valence bond theory.
- Atomic orbitals can combine, or hybridize.
- The bulk allotropes of carbon are diamond, graphite, and amorphous carbon.
- For thousands of years it was believed the only three forms of pure carbon that existed were the bulk allotropes. In the 1980's a new form of pure carbon was discovered – the Buckminsterfullerene, or closed cage of carbon.
- After the discovery of the Buckminsterfullerene, other nanostructures of pure carbon were discovered, namely carbon nanotubes and graphene.
- Both carbon nanotubes and graphene have unprecedented physical properties and have potential for use in a host of applications.

REFERENCES

1. J. McMurry, *Organic Chemistry*, *8th Ed.*, Brooks Cole, 2011.
2. H. Kroto, J. Heath, S. O'Brien, R. Curl, R. Smalley, "C_{60}: Buckminsterfullerene," *Nature*, **318**, *1985*, p.162.
3. C. Bingel, "Cyclopropanierung von Fullerenen," *Chemische Berichte*, **126 (8)**, 1957, p. 1957.
4. W. Mickelson, S. Aloni, W. Han, J. Cumings, and A. Zettl, "Packing C_{60} in Boron Nitride Nanotubes," *Science*, **300 (5618)**, 2003, p. 467.
5. R. Bolskar, A. Benedetto, L. Huesbo, R. Price, E. Jackson, S. Wallace, L. Wilson, and J. Alford, "First Soluble $M@C_{60}$ Derivatives Provide Enhanced Access to Metallofullerenes and Permit in Vivo Evaluation of $Gd@C_{60}[C(COOH)_2]_{10}$ as a MRI Contast Agent, " *J. Am Chem Soc.*, **125 (8)**, 2003, p. 5471.
6. S. Iijima, "Helical Microtubules of Graphitic Carbon," *Nature*, **354,** *1991*, p. 56.
7. R. Booker and E. Boysen, *Nanotechnology for Dummies*, *1st Ed.*, For Dummies, 2005.

8. S. Trans, A. Verschueren, C. Dekker, "Room-temperature Transistor Based on a Single Carbon Nanotube," *Nature,* **393,** *1998,* p. 49.

9. K. Novoselov, A. Geim, S. Morozov, D. Jiang, Y. Zhang, S. Dubonos, I. Grigorieva, A. Firsov, "Electric Field Effect in Atomically Thin Carbon Films," *Science,* **306 (5696),** *2004,* p.666.

CHAPTER 6   NON-CARBON NANOMATERIALS

INTRODUCTION

While carbon-based nanomaterials are widely used and very well characterized, there many other types of nanoscale materials with equally interesting properties and potential applications. Chapter 6 will outline some of these other types of nanoparticles and compare them with their bulk counterparts. The chapter begins by describing several metallic nanoparticles, namely gold, silver, platinum, palladium, rhodium, and cobalt. There are also some ceramic nanomaterials that will be discussed such as iron oxide, silicon dioxide, and titanium dioxide, as well as quantum dots, or semiconductor nanoparticles. You will learn about the exciting potential these materials hold for future applications based on their unprecedented properties.

Some things to think about:
- How do nanoparticles of materials such as silver and gold behave differently from bulk materials?
- Why is it surprising that nanoparticles of metals such as platinum make excellent catalysts?
- How are cobalt and iron oxide nanoparticles similar and how do they differ?
- What are some unique properties of quantum dots? How might you imagine they can be used?
- What are some other bulk materials that might make interesting nanoparticles?

6.1   GOLD

The uniqueness of gold nanoparticles (**colloidal gold**) compared to their bulk counterparts was briefly discussed in Chapter 4. It is easy to picture bulk gold that is yellow, shiny, and a good conductor of electricity. Gold nanoparticles, on the other hand, are quite different. Their color will depend on the diameter of the nanoparticles, and can range across the visible spectrum from red to violet. As opposed to being good metallic conductors, gold nanoparticles are semiconducting.

Because of their distinct interaction with visible light, it was mentioned that gold nanoparticles have been used for centuries, for example, to color glass. More recently, these unique properties have been studied and used in very high technology applications. The oscillating electric fields of a ray of light originating near a gold nanoparticle will interact with the gold's free electrons. The free electrons can then also oscillate in resonance with that particular frequency of visible light. These resonating electron oscillations are **called surface plasmons**.

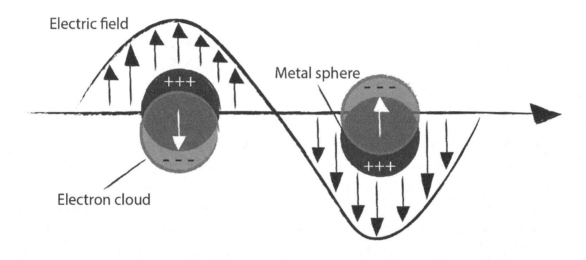

FIGURE 6.1

Figure 6.1 Surface plasmon resonance

For a solution of smaller gold nanoparticles that are on the order of 10 nm, the surface plasmon resonance phenomenon causes light to be absorbed in the blue to green portion of the visible spectrum, around 450 nm. When this portion of the spectrum is absorbed, the longer red wavelengths of light, around 700 nm, are reflected. The result is a solution that is a vibrant red color. As the size of the gold nanoparticles increases, there is a shift in the absorption toward the longer red wavelengths. When red light is absorbed, then blue-violet light is reflected, yielding solutions that are light blue and violet in visible color.

As the size of the colloidal gold particles increases to the point where they are no longer on the nanoscale, the surface plasmon resonance wavelengths move out of the visible light portion of the electromagnetic spectrum and into the infrared. Since we cannot see these wavelengths of light, the colloidal gold solutions will appear colorless and clear to our eyes. Since it is quite easy to synthesize these various sizes of gold nanoparticles, the result is a facile way to specifically tailor optical properties for various applications.

There is a large range of applications for these gold nanoparticles, and new ideas are consistently being generated. Currently, gold nanoparticles are being used for different sensors, both chemical and biological, including DNA detection and detection of air pollutants. There is also interest in using gold nanoparticles in electronic devices. They can be printed onto surfaces an "ink," or used as contacts in chips to connect different elements like resistors and conductors. Other potential applications include use in microscopy probes, diagnosis of infectious agents, and catalysis of chemical reactions.

Gold nanoparticles have a lot of promise in the field of medicine, which will be further discussed in subsequent chapters. Gold nanoparticles are easily functionalized with organic molecules, since it is well known that elements such as sulfur readily bond to their surface. It is possible to attach therapeutic agents to the surface of gold nanopartucles and coat their entire surface. The larger gold particles that absorb light in the red to infrared range (700 to 800 nm) actually produce heat upon absorption. The idea is that this property can be exploited to "cook" tumor cells in a process known as hyperthermia therapy.

6.2 SILVER

Another system of metallic nanoparticle with many current and potential applications is silver. Similar to gold, silver nanoparticles, or **colloidal silver**, also have very unique optical, electrical, and thermal properties. The surface plasmon resonance phenomenon observed with gold nanoparticles also occurs with silver. Their optical properties can easily be tuned as well. When smaller silver nanoparticles in the 50-60 nm range interact with white light, they absorb the longer red wavelengths and appear bright blue. The bright blue is a result of a surface plasmon resonance that peaks at 450 nm. Again, by changing the size of the silver particles, the surface plasmon resonance peak wavelength can be tuned from 400 nm for violet to 530 nm for green. By changing the shape of silver nanoparticles to rods or plates, the surface plasmon resonance can be adjusted to peak in the infrared region, and heat may be produced upon absorption.

The surface of silver nanoparticles can also easily be functionalized with various molecules. It is known that sulfur and nitrogen bond to silver's surface, allowing for relatively easy functionalization with simple molecules, or even larger molecules like polymers, antibodies, or proteins. The ease of functionalization makes silver nanopartices good candidates for use as biosenors, or biological tags, in disease detection. Because they have excellent thermal and electrical conductivity, silver nanoparticles could potentially be used as conductive links in electronics, or could be added to materials to make composites with better thermal and electrical conductivity.

One property that makes silver nanoparticles stand out from others are their unique ability to kill bacteria. Because nanoparticles have such a large surface area-to-volume ratio, there are many silver ions at their surface which are known to kill bacteria. As far back as ancient Greece, humans have been exploiting silver for its antimicrobial properties.

The Greek physician Hippocrates who lived around 400 BC, also known as the "father of Western medicine," noticed that silver had healing properties against disease.

The ancient Phoenician civilization who lived in the area around the Mediterranean Sea from around 1550 BC - 300 BC stored their water, wine, and vinegar in bottles made of silver to prevent anything from spoiling. Since its germicidal effects were known, silver's value was high, especially in use as utensils and in jewelry. In World War I, before the advent of antibiotics, silver was used in topical creams to prevent wound infection.

Colloidal silver continues to be used, and can be found embedded in many consumer products to kill bacteria. Silver nanoparticles have been incorporated into such products as clothing, shoes, socks, bandages, hospital linens, cosmetics, plastic food storage containers, and paints. They have also been added to filters to aid in the cleaning of contaminated drinking water.

6.3 PLATINUM

Platinum is a heavy metal, element #78 on the periodic table. Bulk platinum is in high demand for use in jewelry because it is the least reactive naturally occurring metal. Aside from being the least reactive metal, it is also one of the rarest elements on earth, and as such is very expensive. Aside from jewelry, platinum is often used in automobile catalytic converters. This **catalyst** helps aid the reaction of converting dangerous gases such as carbon monoxide into safer air emissions.

Platinum nanoparticles have received much attention due to their antioxidant properties. An **antioxidant** is a material that prevents other molecules from oxidizing. Oxidation of molecules produces highly reactive **free radicals**, molecules which contain an unpaired electron. Since we know that molecules want to have full pairs of electrons, a free radical will remove an electron from another species, creating a new free radical, and as such, an entire chain reaction. Free radicals may be responsible for the onset of many diseases, including Parkinson's, Alzheimers, arthritis and some forms of diabetes.

Much research has demonstrated the use of platinum nanoparticles as effective antioxidants, including inhibition of pulmonary inflammation from cigarette smoke, and reduction of bone loss after estrogen deficiency.

Platinum nanoparticles are also effective catalysts for chemical reactions. It has been shown that the ability to control the size and shape of platinum nanoparticles will help to control their catalytic activity. They have also been used in various applications such as sensing biological molecules, and cancer treatment therapies.

6.4 PALLADIUM AND RHODIUM

Palladium and rhodium are other rare metals that have properties similar to platinum, but cost about half as much as platinum. Both metals are also used in catalytic converters as well as for jewelry. Nanoparticles of palladium and rhodium are more effective in catalytic converters than their bulk counterparts due to their larger surface area, and are more cost effective since significantly less material is required.

The usefulness of palladium and rhodium as cost effective substitutes for platinum as catalysts for chemical reactions has been well established. It has been found that, similar to platinum, the controlling the morphology of these nanoparticles helps to dictate their catalytic abilities.

6.5   COBALT

Cobalt, as was mentioned in Chapter 3, is a ferromagnetic metal that is only found in cobalt compounds in nature. Cobalt compounds have been used for centuries, due to their rich and vibrant colors. Since the Bronze Age, these compounds have been used to color glass, with samples of blue cobalt-colored glass dating as far back as the 14th century BC. Cobalt-based colorants have been found in Egyptian sculptures, in Italian ruins at Pompeii, and in China in porcelain pieces.

Cobalt nanoparticles have attracted attention because, like bulk cobalt, they are magnetic. There is interest in using cobalt nanoparticles as a **contrast agent** for **magnetic resonance imaging** (MRI). By adding magnetic nanoparticles like cobalt into cells, they can be tracked inside the body using MRI. The current contrast agents used are made using the element gadolinium, which as discussed previously, can have harmful side effects. Other types of nanoparticles cannot be as magnetized as cobalt, and therefore are not as attractive as candidates to replace gadolinium agents.

In addition to medical uses, magnetic nanoparticles like cobalt could be used in the field of information and data storage. The idea of storing information via magnetic materials is not a new one. Audio recordings were first stored on magnetic media in the late 1800s. The first magnetic recorder was invented in 1898, and the first magnetic tape recorder was invented in 1928 by German engineer Fritz Pfleumer.

Audio and video digital data is now mostly stored via magnetic devices. **Magnetic storage** media use different patterns of magnetization to store data in what we know as non-volatile memory, or computer memory that can keep information even when your computer is turned off. As we want to store more and more information on our electronic devices, we need to continue to scale down memory components so the devices can remain small and lightweight. Bulk ferromagetic material has reached the limit of how small it can be and still be useful, so magnetic nanoparticles could be used to be able to continue shrinking our devices.

One major drawback to cobalt nanoparticles is their toxicity since there is potential for cobalt poisoning. Cobalt poisoning occurs when there is cobalt levels in the body that are too high. While small amounts of cobalt are necessary for life, high levels can lead to pernicious anemia, which is often fatal.

6.6 IRON OXIDE

In addition to the many metallic nanoparticles being used in a wide variety of applications, many ceramic nanomaterials are also of interest. Perhaps the best characterized of the ceramic nanoparticles is iron oxide. Like cobalt, iron is also magnetic, and so iron ceramic compounds such as iron oxide are also magnetic.

There are two main forms of iron oxide nanoparticles: Fe_3O_4, which is often called magnetite, and Fe_2O_3, or maghemite. Both Fe_3O_4 and Fe_2O_3 received a great deal of attention, mainly because they are superparamagnetic and far less toxic than cobalt.

Superparamagnetism is a special form of magnetism that occurs only in small nanoparticles that are ferromagnetic. In Chapter 3, ferromagnetic materials were defined as having all of their atom magnets pointing in the same direction. You can still have pieces of ferromagnetic material that are unmagnetized. This is possible because materials are divided up into magnetic domains that have all of their atom magnets pointing in the same direction. While each domain has all of its atom magnets pointing in the same direction, the electrons of different domains point in different directions and can have an overall canceling out effect. The result is a material that is not magnetic. In superparamagnetism, the nanoparticles are made up of one single domain, instead of multiple domains that can cancel out. What this means is that each iron oxide nanoparticle has a fixed magnetic moment, and the magnetic moments do not cancel.

There are many potential applications for magnetic iron oxide nanoparticles. As with cobalt, there is interest is using such magnetic nanomaterials for magnetic information storage media and MRI contrast agents. Iron oxide also has potential in biomedical applications. Iron oxide is biologically compatible with humans and relatively non-toxic. Much research has used iron oxide nanoparticles as targeted drug delivery agents. While there have been many advancements in cancer treatments, cancer still remains a leading cause of human death. Current treatments include surgery, radiation therapy and chemotherapy. Surgical removal of tumors is challenging since one must not damage any of the surrounding normal tissue while removing as many cancer cells as possible. Chemotherapy is systemic and highly invasive, and often does not expose a tumor to high enough concentrations of anticancer drugs. There is poor selectivity of these chemotherapy treatments toward tumor cells and can harm other normal, healthy cells, which contributes to the negative side effects of these drugs.

Nanotechnology may begin to change the way in which cancer is treated. Iron oxide, for example, may act as a carrier of anti-cancer agents. The nanoparticles are relatively easily functionalized, and may be brought to specific tumor sites via an external magnet. Their location can also be easily visualized via MRI. The result is a therapy which is much more targeted and far less invasive than current treatments. The details of this process will be discussed in Chapter 8.

6.7 QUANTUM DOTS

Quantum dots can be defined as nanoparticles of a semiconducting material, for example cadmium selenide and indium phosphide. They were discovered in the 1980s by Russian solid state chemist Alexei Ekimov and then in colloidal solution by American chemist Louis E. Brus.

These semiconductor nanoparticles are of particular interest because they have very unique electronic and optical properties. As mentioned in Chapter 3, in order for a semiconductor to conduct electricity, the electrons in the valence band must be promoted into the conduction band. In quantum dots, this process is confined in all three dimensions to a nanoscale particle which results in properties that are somewhere between a bulk semiconductor and an individual molecule. The band gap of quantum dots of the same material is inversely proportional to the size of the nanoparticle. In other words, the smaller the quantum dot is, the larger its band gap. This means that in smaller particles more energy is required to promote electrons across the band gap, so more energy is released upon relaxation back to the ground state. The result is emission of higher frequencies of light. As the size of quantum dots decreases, the resulting color of light emitted by the dots shifts from red to blue.

Due to their easily tunable optical and electrical properties, there are many potential applications for quantum dots. On such application is their use in medical imaging. Currently, various organic dyes are being used to visualize biological materials such as specific cells. Quantum dots have been found to be far superior to these organic dyes as they are much brighter, easier to visualize, and have a longer lifetime. They can be used to image, for example, specific pre-labeled cells in vitro and in real time, which will aid in such fields as cancer cell research and stem cell therapies. There have also been recent attempts to use quantum dots for in vivo (in the body) for targeting tumors. The quantum dots are functionalized with tumor specific binding molecules that will selectively bind to tumor cells only. There is some concern with potential toxicity; however, there is still more research to be done concerning the health effects of quantum dots.

Quantum dots could also be used for new high definition displays. Several groups have used quantum dots as light-emitting diodes, or "QD-LED" displays. The current technology is generally dominated by liquid crystal displays (LCDs), which are relatively inefficient in their energy consumption. Organic light emitting diode (OLED) displays have very vibrant colors, but there have been issues with molecule stability and cost for large area displays. As an alternative, some companies have been exploring the use of quantum dots to enhance LCDs. As we know, quantum dots emit brightly when their electrons are excited and relax back to ground state, which would lead to the production of displays with rich and lively colors. They would be relatively inexpensive to manufacture.

Finally, quantum dots could potentially increase the efficiency of solar cells. Currently, solar cells are made from silicon so are bulky, expensive, and somewhat inefficient. Again, quantum dots are relatively inexpensive to manufacture, are small and could produce lightweight and more efficient devices.

6.8   SILICA

Silica is a material made of a silicon atom bonded to two oxygen atoms, SiO_2. Silica is not to be confused with silicone, which is a polymer of silicon atoms. In nature, the most common bulk forms of silica are sand and quartz. Bulk silica has many uses, including fiber optic wires, ceramic dishware, cement, and as an additive in the production of foods. Silica is also converted into electronic grade silicon in semiconductor manufacturing.

There are several forms of nanoscale silica that have a wide range of potential applications. **Colloidal silica** is an aqueous suspension of silica nanoparticles that may range in size from 30-100 nm. One major application of colloidal silica is in the semiconductor manufacturing industry. Aqueous slurries of silica nanoparticles are used as abrasives to polish silicon wafers. These same slurries are also often used for polishing optics and other glass components. Colloidal silica can also be used to catalyze, or lower the activation energy, of chemical reactions. Its applications also include abrasion-resistant coatings, absorption of moisture, and as high friction agents to improve the traction of floors, fibers, and tracks.

Another nanoscale form of silica is known as **mesoporous silica**, or silica nanoparticles that have pores within them. The term mesoporous refers to a material that has small pores on the order of 2-50 nm. A material with smaller pores would be considered microporous, while a material with larger pores is classified as macroporous. Mesoporous silica's pores have a very large surface area, and as such have great potential to act as a drug delivery vehicle. The silica's pores can be impregnated with a drug, and when functionalized properly, a cell will take in these small porous particles. Once in a cell, the drug can diffuse out of the mesoporus silica nanoparticles. The pores also allow for these particular types of silica nanoparticles to be filled with different dyes that might not otherwise be allowed across a cell membrane. These dye-filled mesoporous silica nanoparticles can then be used as sensors for specific biomolecules within a cell.

A third silica nanostructure is silica **aerogel**. An aerogel is a type of solid foam, like the more familiar Styrofoam™. Since aerogels are made of gas dispersed in solid, they are porous and incredibly lightweight. The material comes from a gel, which is a liquid dispersed in a solid, but in the case of aerogel, the liquid component is dried and replaced with gas through a process known as supercritical drying.

Silica aerogels were first synthesized by American chemical engineer Samuel Stephens Kistler in 1931, who figured out how to remove the liquid component from gels without damaging the structural integrity of the solid. Through the process of supercritical drying, the liquid component of the gel can be slowly dried without the solid silica collapsing. The result is a solid, dry, rigid material with spherical nanoscale pores that can bear a comparably heavy load. Silica aerogel is an incredible thermal insulator since it is made mostly of gas, which is not a good conductor of heat. Silica aerogel will not melt until 1,200 °C, and has many entries in the Guinness Book of World Records for various material properties, including the best thermal insulator and lowest density solid.

There are many potential applications for such an interesting material. Since the nanoscale pores have such a large surface area, silica aerogel could be used as an adsober of chemicals, for example cleaning chemical spills, or for carrying a chemical catalyst. Because it is such a good thermal insulator and can withstand high temperatures, silica aerogel could also be used as a thermal insulator for windows and for insulation in aerospace applications such as in space suits and on space shuttles. There is also interest in making suits of silica aerogel to protect Navy divers from cold temperatures.

6.9 TITANIA

Titania is also known as titanium dioxide, or TiO_2. It is a naturally occurring ceramic material that has many applications. The most common application of titania, both bulk and nanoscale, is as a white pigment. Titanium dioxide has one of the highest known refractive indices of any material and as such appears a very bright white. Titania can be found in paints, inks, foods, cosmetics, medicines, and toothpaste. Another way to exploit titania's unusually high refractive index is in sunscreens. Almost all commercially available sunscreens contain titanium dioxide nanoparticles as a blocker of ultraviolet light.

Titania is also often used as a catalyst for chemical reactions that involve light (a photocatalyst). This particular property of titania was first discovered in 1967 by Japanese chemist Akira Fujishima. As will be discussed in Chapter 9, there is a need to create hydrogen for use in hydrogen fuel cells. Titania has the ability to break water down into hydrogen and oxygen in a process known as hydrolysis. Titania has also been used as a photocatalyst in a type of solar cell that uses organic dyes, also known as a **Grätzel cell**.

Another special property of titania is unusually high **capacitance**, or ability to hold charge, known as a dielectric constant (κ). The dielectric constant of a material is the ratio of the amount of energy stored in a material relative to that stored in a vacuum, and as such has no units. Glass, for example, has a dielectric constant of around 2-5, while titania's dielectric constant ranges from 86-173. There is interest in using TiO_2 nanoparticles for large area displays, as they can hold charge well and would not deplete battery life quickly.

NANOSPEAK – WITH SIMON MINER (CHAPTER 6)

1. Where do you work and what is your job title?
    Optics and Optographics Sales Engineer at Gurley Precision Instruments.

2. What is your educational background?

I have AAS in Nanoscale Materials Technology and am still working on BBA in Business and Technology Management at Delhi University.

3. How did you get interested in nanotechnology/where did you hear about it?

I got interested in nano in particular because of a Time Magazine special in '99 talking about what would be hot in the 21st century. I obviously always had a love of science and tech though.

4. What advice would you give to someone who might want to get into the field?

My advice to students trying to enter the field is not to forget that there are opportunities outside the semiconductor world they should not ignore some of the smaller support companies. There are companies that are using different forms of nanotechnology that they might find even more interesting!

CHAPTER 6 SUMMARY:

- Chapter 6 outlines various metallic and ceramic nanoscale materials.
- Gold nanoparticles are very different from bulk gold. They are semiconducting, and their optical properties can be tailored by changing particle size and shape.
- Gold nanoparticles' free electrons can interact with light via a phenomenon known as surface plasmon resonance.
- Some applications for gold nanoparticles include catalysis, medical diagnostic, and drug delivery.
- Silver nanoparticles also exhibit the surface plasmon resonance phenomenon, with tunable optical properties based on particle size and shape.
- Silver nanoparticles are known to have antimicrobial properties and can be used in a number of applications to kill bacteria.
- Platinum, palladium and rhodium are all rare, expensive and unreactive bulk metals. Nanoparticles of these metals, however, are reactive and useful as catalysts for chemical reactions.
- Cobalt and iron oxide are magnetic both in bulk and in nanoscale form. They are being researched for use in medical imaging, digital information storage, and targeted drug delivery.
- Quantum dots are nanoparticles of semiconducting materials such as cadmium selenide or indium phosphide. They have exceptional optical and electrical properties wherein their band gap and color can be adjusted simply by changing the particle size.
- Silica, or silicon dioxide, exists in several nanoscale forms – colloidal, mesoporous, and aerogel. Colloidal silica can be used for polishing in an aqueous slurry, mesoporous silica is often impregnated with therapeutic agents, and aerogel is best known for its low density and high thermal insulation.

- Titania, or titanium dioxide, has a high refractive index and appears bright white. It is used for paints, pigments and cosmetics. Many sunscreens contain titania nanoparticles because they block ultraviolet light. Titania also makes an excellent catalyst for reactions that use light.

REFERENCES

1. S. Link and M. El-Sayed, "Size and Temperature Dependence of the Plasmon Absorption of Colloidal Gold Nanoparticles," *J. Phys. Chem. B,* **103 (21),** *1999,* p. 4214.

2. X. Qian, "In Vivo Tumor Targeting and Spectroscopic Detection with Surface-Enhanced Raman Nanoparticle Tags," *Nature Biotechnology,* **26 (1),** *2008,* p. 83.

3. R. Booker and E. Boysen, *Nanotechnology for Dummies, 1ˢᵗ Ed.,* For Dummies, 2005.

4. M. Yin, A. Willis, F. Redi, N. Turro, S. O'Brien, "Influence of Capping Groups on the Synthesis of γ-Fe$_2$O$_3$ Nanocrystals," *Journal of Materials Research,* **19 (04),** *2004,* p.1208.

5. J. Alexander, "History of the Medical use of Silver," *Surgical Infections,* **10 (3),** *2009,* p. 289.

6. J. Kim, "Effects of a Potent Antioxidant, Platinum Nanoparticle, on the Lifespan of C. Elegans," *Mechanism of Ageing and Development,* **129 (6),** *2008,* p. 322.

7. L. Parkes, R. Hodgson, "Cobalt Nanoparticles as a Novel Magnetic Resonance Contrast Agent," *Contrast Media Mol Imaging,* **3 (4),** *2008,* p. 150.

8. L. Brus, "Electron-Electron, and Electron-Hole Interactions in Small Semiconductor Crystallites," *Journal of Chemical Physics,* **80,** *1984,* p. 4403.

9. S. Kistler, "Coherent Expanded Aerogels and Jellies," *Nature* **127,** *1931,* p. 741.

10. A. Khatee, *Nanostructured Titanium Dioxide Materials: Properties, Preparation, and Applications,* World Scientific Publishing Company, 2011.

CHAPTER 7   Synthesis and Characterization at the Nanoscale

INTRODUCTION

The design and manufacture of nanoscale materials is known as nanofabrication. Nanofabrication can occur via either a top-down method or bottom-up method. Top-down manufacture refers to methods by which nanoscale structures and features are formed in a bulk material by a series of steps that put down films in patterns and etch away certain parts of the films. The steps involved in top-down nanofabrication are deposition, etching, modification, and lithography.

Bottom-up nanofabrication makes nanostructures from individual atoms or molecules. There are several methods for bottom-up nanofabrication, including colloidal synthesis, thermal decomposition, and sol-gel synthesis.

Since our eyes cannot see materials on the nanoscale, it is important to understand how we 'see' and characterize nanostructures. You will learn about the various forms of microscopes we can use to visualize nanoscale materials, as well as some methods of spectroscopy that help gives us information about the chemical make-up of nanomaterials.

Some things to think about:
- What is the difference between bottom-up and top-down nanofabrication?
- Why would it be ideal to use bottom-up nanofabrication for the manufacture of semiconductor devices?
- Why are electron beam microscopes used for visualizing nanoscale materials?
- How do X-rays give us clues about the chemical make-up of materials?

7.1   AN INTRODUCTION TO NANOFABRICATION

In the field of nanotechnology, **nanofabrication** is the process by which nanostructures are made. There are a wide variety of materials that can be created such as nanotubes and nanowires, nanoscale thin layers (planar structures) and hybrid structures, which are a mixture of both nanoparticles and thin films.

There are two methods by which nanofabrication can occur. The first is known as top-down, and the second is known as bottom-up. There is also hybrid nanofabrication, which combines elements of both top-down and bottom-up techniques. Top-down nanofabrication refers to procedures by which nanostructures are formed in a bulk material by a series of steps that put down thin films and etch certain parts away. Bottom-up nanofabrication builds up nanostructures from atoms, molecules, and/or individual particles. You can use the analogy of top-down fabrication being likened to carving a sculpture out of a rock, and bottom-up fabrication like putting individual blocks together to form a sculpture.

7.2   TOP-DOWN NANOFABRICATION

Top-down fabrication generally begins with an initial material, or substrate, onto which new materials can be added and removed according to a very specific pattern. There are four basic steps that are used in top-down nanofabrication:

1. Deposition
2. Etching
3. Modification of Materials
4. Lithography

These four steps may be used in any sequence that can be repeated multiple times, and not all steps may need to be used. In general, the sequence of steps begins with deposition on a thin layer of material on to a substrate, and all other steps are directed by lithography, which dictates where materials stay, get etched away, and get deposited in the nanosculpture.

1.1.1   DEPOSITION

Deposition is also known as material growth. It is required in nanofabrication to create the basic layers from which nanostructures are formed. There are several methods by which deposition can be accomplished. The first is growth by chemical reaction, like via oxidation. In the semiconductor industry, wafers of pure silicon are used as the initial substrate. A thin film of silicon dioxide can be grown atop the silicon to act as an insulator. This process can generally be accomplished via a chemical reaction between the silicon wafer and oxygen gas in a specialized high temperature furnace via the following scheme:

$$Si_{(solid)} + O_{2\,(gas)} \rightarrow SiO_{2\,(solid)}$$

The ability to oxidize a silicon substrate with relative ease and form an insulating SiO_2 thin film layer is one of the main reasons silicon is the material of choice in the microelectronics industry.

Another method by which deposition of thin film layers can be accomplished is by physical application of a material onto a substrate. There are many methods by which thin films can be physically applied onto a substrate like dipping, spraying, or spin coating. In the process known as **spin coating**, an excess amount of the material is added atop the substrate in solution in a machine known as a spin coater. A spin coater is an instrument that contains a quickly rotating plate that holds the substrate via a vacuum. Once the desired solution is applied, the chuck inside the spin coater rotates while the excess material spins off the edges of the substrate, leaving behind a thin film. With the film in place, the solvent that had dissolved the desired material can evaporate off, leaving behind a solid film.

In the microelectronics industry, spin coating is often performed with a unique material called **photoresist**. Photoresist is a light-sensitive polymer that is very important in the lithography step of nanofabrication.

Physical vapor deposition (PVD) and **chemical vapor deposition** (CVD) are two other ways in which growth of thin films may be achieved. Physical vapor deposition is a broad term used to describe a variety of different methods of thin film deposition wherein a vaporized form of the desired thin film material is condensed onto a substrate. There are many forms of PVD, including sputter deposition, and evaporative deposition.

In **evaporative deposition**, atoms are transferred from a target to the substrate located some distance away using heat. The distance will depend on the chamber size. Enough heat must be imparted to the target, or material to be deposited, so that the solid target material will effectively evaporate or sublime. Sputter deposition, by contrast, uses an ionized gas, or plasma, to bombard the target material, which then causes target atoms to vaporize and deposit onto the substrate. In both examples, there are no chemical changes that occur, only physical phase changes.

Chemical vapor deposition (CVD), by contrast, involves a chemical reaction in order to produce thin films. In CVD, process gases are introduced into a reaction chamber, undergo a chemical reaction on the substrate's surface and deposit a solid product on the substrate. There are gaseous byproducts of the reaction that leave the substrate's surface and are vented from the chamber. CVD is a process that requires several steps, outlined below.

1. Process gases are introduced into a reaction chamber.
2. Reaction precursors reach the surface of the substrate.
3. The reaction precursors adsorb, or adhere, to the substrate's surface.
4. There is a process that occurs by which these reaction precursors migrate along the substrate's surface.
5. The chemical reaction at the surface of the substrate begins.
6. The solid byproducts from the chemical reaction leave crystal nuclei or tiny seed crystals.
7. The seed crystals begin to grow into larger crystals called islands.
8. The islands continue to grow and eventually meet up and become one continuous thin film.
9. Other byproducts of the chemical reaction, generally gases, come off of the substrate.
10. The byproduct gases flow out of the reaction chamber.

1.1.2 LITHOGRAPHY

In top-down nanofabrication it is important to use some type of pattern to direct which areas on a substrate have films deposited on them, which area need to be etched away, and which areas need further modification. This pattern needs to be transferred onto the substrate via a process known as **lithography**. The word lithography comes from the Greek 'litho,' which means stone, and 'graphy,' which means to write.

Lithography begins with designing a pattern onto what is known as a mask or reticle. In the semiconductor industry, for example, the circuit design pattern to be made on a microelectronics chip needs to first be generated onto a silicon wafer. This process requires the circuit design be made via electronic design automation software and printed onto a piece of chromium-coated quartz glass. This printed piece of quartz glass serves as the director or template of the pattern to be transferred onto the silicon wafer, and is known as a mask. When this chromium-coated quartz glass covers only a part of the entire substrate it is called a reticle.

The most common form of lithography used in the fabrication of nanoscale materials is known as **photolithography**. In photolithography, light-sensitive polymers are used to pattern a substrate. As mentioned previously, these special light-sensitive polymer materials are known as photoresists. Photolithography is of utmost importance in the semiconductor manufacturing industry. A microelectronics chip may require more than 30 different mask and photolithography steps to be complete.

Photolithography begins with cleaning the substrate via a chemical cleansing method, followed by a two-step substrate preparation procedure. The first step involves heating the substrate in order to remove any moisture on the surface. This first step, called the prebake or the dehydrate bake is essential for allowing the photoresist to stick to the substrate's surface effectively. The second step is known as priming. In this step, a thin layer of a primer material that helps the organic photoresist adhere to the silicon substrate is coated onto the silicon substrate's surface. The primer is vaporized, introduced into the process chamber, and deposited onto the wafer's surface with the prebake step.

In order to prevent water from adsorbing onto the silicon substrate, the photoresist generally needs to be applied immediately after the prebake and priming. Coating the substrate in photoresist is a deposition process, as outlined in 7.2.1. Again, the silicon substrate is placed on a spinning chuck that holds the wafer with vacuum force during high speed spinning. Liquid photoresist is added onto the substrate's surface and the spinning helps to spread the photoresist evenly across the wafer. Thickness of the photoresist layer may be controlled by spinning rate and photoresist viscosity.

After the photoresist has been coated onto the silicon substrate, the wafer is put through a process known as the soft or pre-exposure bake, which removes any excess solvent that is in the photoresist mixture. With a baked on layer of photoresist in place, the wafer is ready for the most crucial part of photolithography: the alignment and exposure steps.

The alignment of the coated silicon substrate and exposure to ultraviolet light are the steps that will transfer the circuit design from the mask or reticle onto the substrate's surface. Exposing the wafer to light is somewhat similar to taking a photo with a film-containing camera. In the same way that an image would be exposed onto photographic film inside a camera, the pattern from the mask or reticle is exposed to the photoresist on the wafer. As opposed to a camera, however, the resolution of the mask or reticle exposure is incredibly high, since the features to be printed are on the nanoscale. The alignment must also be incredibly precise, otherwise the pattern will not successfully transfer to the wafer's surface, and the chip will not work.

In general, a light source passes through a projection lens, through the mask or reticle, and through a second lens which focuses the pattern onto the photoresist. There are two types of photoresist: positive and negative. In the case of negative photoresist, the areas of the coated substrate that get exposed to light become cross-linked and polymerize due to a photochemical reaction with the light. These areas harden and remain on the wafer surface after development. The areas that were not exposed to light will rinse off during development. For positive photoresist, the polymer is already cross-linked. After exposure to light, the uncovered areas break down and will dissolve during development, while the areas that were not exposed to light remain on the surface of the substrate. Positive photoresist is most commonly used in the semiconductor manufacturing industry.

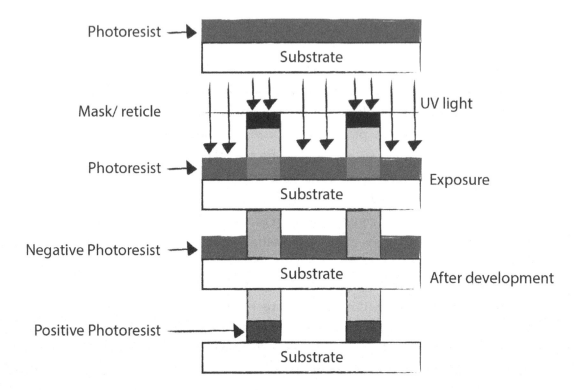

FIGURE 7.1

Figure 7.1 Positive vs. negative photoresist

Following exposure of the coated substrate, there is another baking process called the post-exposure bake process. The photoresist-coated wafer then goes to the developer station. First, the wafers are rinsed in a developer solution, to rinse away the areas of photoresist that were exposed to light. The most commonly used developer of positive photoresist is tetramethyl ammonium hydroxide. The wafers are then rinsed with water and spun dry. After the development process, the wafers are put through a hard bake process to remove any remaining solvent in the photoresist and strengthen it. The last step in photolithography is pattern inspection to ensure the pattern has successfully been transferred onto the wafer.

In addition to photolithography, there are other forms of lithography used in nanofabrication, including **dip pen lithography** wherein an atomic force microscope probe tip is used to write on a substrate. Another lithography method is called embossing or **nano-imprinting**. In nano-imprinting, a mold with nanoscale features is pressed into a layer of resist coated onto a substrate. An additional lithographic technique is called soft lithography or **microcontact printing**, which involves the transferring of a pattern onto a substrate via a patterned stamp. In **electron beam lithography**, or e-beam lithography, a pattern is drawn into a resist using a beam of electrons. E-beam lithography does not require a mask or reticle for pattern transfer to resist, but can take a long time to complete.

1.1.3 ETCHING

Etching is an important process in nanofabrication that removes materials from a substrate's surface. Etching can either be patterned, where only selective areas of materials are removed from a designated design, or blanket etching, wherein all or part of the surface film is removed. The etching process uses wet chemical etching, dry physical etching, or dry chemical/physical (reactive ion) etching.

Wet etching is a process that uses a chemical solution to remove material from a substrate's surface in three basic wet etch, rinse, and dry steps. Wet etch was more popular in the semiconductor manufacturing industry before the 1980's, when the size of chip features was larger than 3·m. It is not effective for smaller feature sizes, as it etches equally in all directions and will undercut the mask pattern deposited onto the substrate. This equal direction etching is also known as **isotropic etching**.

Dry, or **plasma etching**, uses gaseous chemical materials to etch away areas of a coated wafer. A plasma is an ionized gas that contains a mixture of neutral gas molecules, charged gas molecules, and electrons. There are various collisions that take place within the plasma, including ionization, excitation-relaxation, and dissociation. Without these collisions, the plasma would not be sustained. These collisions also generate highly reactive free radicals that help to increase chemical reactivity for etching.

The ions created in plasmas can bombard the wafer and physically remove materials and also break chemical bonds on the surface. This plasma etching process has generally replaced wet etching processes to more effectively generate nanoscale features. While wet etching is an isotropic process, dry plasma etching is an **anisotropic process**, etching in only one direction.

A final method of etching combines chemical and physical etching methods. In this dry chemical plus physical etching process, plasmas are still used, but the etching properties can be tailored by adjusting the ion bombardment. This method is also known as **reactive ion etching** or RIE. Currently, it is the most prevalent method of etching used in the semiconductor industry, as the degree of anisotropy and selectivity can best be controlled simply by changing the process gas composition and power added to the plasma.

1.1.4 MATERIAL MODIFICATION

The final step of top-down nanofabrication is **material modification**. Material modification refers to any process that is used to specifically tailor chemical, mechanical, optical, or electrical properties on a surface.

One of the most common ways to modify selective areas on a surface is by a process known as **ion implantation**. Ion implantation is the way in which dopants are added to semiconductor materials to tailor their electrical conductivity. Silicon can be doped with trivalent or pentavalent materials to achieve p-type and n-type semiconductors, respectively, the details of which are outlined in section 3.4.3.

Ion implantation is a process that adds the dopant atoms to silicon by forcefully bombarding the surface with a high energy beam of ions. As the ions bombard and enter the surface of the silicon wafer, they collide with the silicon atoms that make up the wafer's crystal lattice. As these collisions occur, the ions lose energy and will stop penetrating the silicon at a given distance into the surface. The collisions of high energy ions with silicon atoms within the crystal lattice cause energy to be transferred from the ions to the silicon atoms. The energy transferred is quite high and can cause atoms to break free from the organized lattice. This damage of the perfect crystal lattice structure can be repaired via an **annealing** process. The annealing process involves exposure of the silicon wafer to high temperatures which restores the silicon to a single crystalline form.

Crystal damage (left) and annealing process

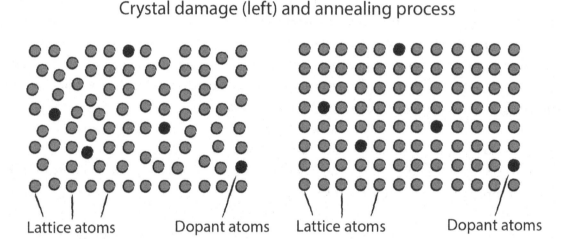

Lattice atoms   Dopant atoms   Lattice atoms   Dopant atoms

FIGURE 7.2

Figure 7.2 The annealing process to restore crystallinity

1.2   BOTTOM-UP NANOFABRICATION

Bottom-up nanofabrication can be likened to building a structure with individual blocks. Small components like atoms, molecules or nanoparticles are assembled into the desired nanoscale structures. There are two basic steps involved in the bottom-up process, the synthesis of the individual building blocks, and then the assembly of the building blocks into functional nanostructures. In bottom-up nanofabrication, there is no use of patterns, lithography or selective etching. Instead, the building blocks self-assemble into structures without patterns directing where they go.

1.2.1 BUILDING BLOCK SYNTHESIS

Since the building blocks in nanofabrication could be anything from atoms or molecules to individual nanoparticles, there are a number of methods by which they may be synthesized.

Often times, **self-assembly** involves individual organic (carbon-based) molecules. These molecules can be tailor-made, depending on if they need to be anchored to a specific surface, or have particular functionality like repelling water. The molecules could also dictate growth of crystalline solids or bind biological molecules like proteins, viruses or DNA. One very common self-assembling system is alkanethiols on gold. A thiol is a molecule that contains sulfur at its end, which readily bonds with metals such as gold or silver. A thiol can be customized with a particular length carbon-chained body, and an end functional group, which can be altered to fit a specific need.

Sometimes, the building blocks for bottom-up fabrication are individual nanoparticles. There are a number of ways in which nanoparticles may be synthesized, one of the most common of which is via a colloidal reaction. A colloid is defined as a microscopic substance evenly dispersed throughout another substance. A colloidal system has two different phases: the dispersed phase particles which are generally on the 1nm to 1·m scale and either solid, liquid, or gas, and the continuous phase, or medium in which the microscopic particles are dispersed.

In **colloidal nanoparticle synthesis**, there is generally an aqueous or organic solution that contains suspended salts that are the precursors to the desired nanoparticles. The salts are then reacted with a second precursor material. Once the solution is saturated with the product of the reaction between the two precursor materials, the nanoparticles start to form and grow.

One type of nanoparticle that is typically synthesized via a colloidal synthesis is cadmium selenide quantum dots. An example of a way in which these particles are made is with a cadmium oxide salt that is reacted with a selenium solution. The reaction occurs in the presence of oleic acid, a molecule that can bind to the surface of the growing nanoparticles and control their size and ability to agglomerate or 'clump' together. The molecules that bind to the surface of nanoparticles are known as ligands.

Another method by which nanoparticles can be made is via a process known as **pyrolysis**, or thermochemical decomposition. In pyrolysis, an organic material is decomposed under high temperature conditions in the presence of a ligand molecule. A common nanoparticle system that is made in this way is iron oxide. An iron-containing organic compound such as iron pentacarbonyl is heated to 350 ºC in a high-boiling temperature organic solvent, such as trioctylamine, with a ligand molecule, commonly oleic acid. The iron pentacarbonyl decomposes to pure iron and is subjected to an oxidant like trimethylamine N-oxide to form iron oxide nanoparticles.

Nanoparticles may also be synthesized by a process known as a **sol-gel synthesis**. Sol-gel synthesis is generally used to make metal oxide nanomaterials. The process begins with a colloidal solution (sol for short) which acts as a precursor for a network (gel) of polymers or individual particles. The precursors are usually metal alkoxides (an organic molecule bonded to a negatively charged oxygen atom) or metal salts. These precursors generally undergo various reactions to form nanoscale materials.

In addition to chemical synthesis, there are also various physical approaches to making nanoscale building blocks. Nanoparticles can be made by a process known as **vapor condensation** whereby metallic or metal oxide nanoparticles can be formed. To make metal nanoparticles, an inert atmosphere is used, and for metal oxides, an oxygenated atmosphere is ideally utilized. Solid nanoparticles are formed when metallic vapor is quickly condensed. This process is often used to make carbon nanotubes. During the vapor condensation growth of carbon nanotubes, a substrate is made with a layer of metal catalyst particles. The metals are typically nickel or cobalt, and sometimes iron. The size of these metal catalyst particles will determine the diameter of the nanotubes that are grown. The substrate is heated to a very high temperature, around 700 ºC, and a mixture of gases are introduced into the reactor. The gases typically include a process gas such as ammonia or nitrogen, and a carbon-containing gas such as ethanol or methane. The carbon-containing gas is decomposed at the surface of the metal catalyst particle, and the carbon migrates to the edges of the particle, where it can form nanotubes.

7.3.2 BUILDING BLOCK ASSEMBLY

Once the individual building blocks have been synthesized for bottom-up fabrication, they need to assemble into the required structures. Self-assembly of building blocks occurs via chemical or physical means.

The example of an alkanethiol was given for molecular synthesis in section 7.3.1. The alkanethiols can self-assemble into one molecule thick layers known as **self-assembled monolayers** (SAMs). The sulfur group will readily anchor to a gold surface, and the end group can be functionalized depending on application. SAMs of alkanethiols on gold are a very facile route to surface modification. The end group may contain, for example, specific antibodies, to which a biological molecule may bind in order to create nanoscale senors. Functionalized alkanethiols can also be microcontact printed onto a gold or silver substrate to control crystal growth.

Nanoscale building blocks may also physically self-assemble. In physical self-assembly, physical forces, such as an electric or magnetic field, or a mechanical force, such as stress, are the driving force for self-assembly. It has been shown, for example, that an applied electric field can cause a solution of carbon nanotubes to self-assemble into a standing up position.

7.4 CHARACTERIZATION TECHNIQUES

Since our eyes cannot physically see any materials at the nanoscale, it is important that we have techniques in order to visualize nanomaterials. There are several ways in which we can obtain information about nanoscale materials. The first is **microscopy**, wherein microscopes are used to actually show us the size, shape, and structure of nanomaterials. There is also spectroscopy, which refers to techniques that give us some information about the chemical composition of materials, as well as some information about chemical and physical properties of nanostructures.

There are different types of microscopes used in science. The traditional microscope most are familiar with is an **optical microscope**, which uses visible light and a series of lenses to give a magnified view of materials. Since the wavelength of visible light is in the range of 400-700 nm, optical microscopy can be limiting when trying to visualize materials at the nanoscale. Even the most sophisticated of optical microscopes cannot be used to give images of samples below the order of hundreds of nanometers.

Most commonly, nanoscale materials can be visualized using electron microscopes. As opposed to visible light, electron microscopes use a high energy beam of electrons to help us see the size, shape, and structure of samples. When the beam of electrons interacts with a sample, a number of different things can happen. First, the electrons may go right through a sample, or get transmitted. Second, some electrons may bounce back off of a sample, known as backscattering. The high energy electron beam may knock some electrons from the sample free, which are known as secondary electrons. Finally, high energy photons (packets of light), or x-rays may be generated from electrons being excited and relaxing back down to their ground state. There are different types of microscopes that take advantage of each of these responses to an electron beam.

A transmission electron microscope (TEM) helps us to see nanoscale materials using transmitted electrons. In order to visualize a sample with a TEM, it must be very thin and allow the electron beam to pass through. Generally, a solution of a sample is evaporated onto a small copper grid before being visualized with a TEM. The TEM can image at much higher resolutions than a traditional optical microscope, even down to the atomic level.

The scanning electron microscope (SEM) uses both backscattered and secondary electrons in order to visualize samples. A high-end SEM can achieve resolution down to the 1 nm scale. Sample preparation for SEM imaging will depend on the instrument. Some samples will be visualized under high vacuum or low vacuum. These samples must be free of water or solvents, and generally are coated in gold or another conductive material if they are not inherently conducting. The coated samples are mounted onto special carbon tape which acts as a ground to prevent the electrons from building up too much charge. There are also environmental SEMs that allow wet or biological samples to be visualized. SEMs produce very high resolution, three-dimensional black, grey, and white images.

Another way in which we can visualize nanoscale materials involves scanning probe tools.

Scanning probe instruments have their name because they contain a small nanoscale tip that scans back and forth across the surface of a sample to give us information about topography. The two most commonly used **scanning probe microscopes** (SPM) in nanoscale materials fabrication are the **atomic force microscope** (AFM) and **scanning tunneling microscope** (STM).

The AFM is an instrument that gives very high resolution surface images, even down to fractions of a nanometer. The first AFM prototype was developed in the 1980s by two IBM researchers, German physicist Gerd Binning and Swiss physicist Heinrich Rohrer.

The way in which an AFM works is by scanning a material's surface with a cantilever that has a sharp tip or probe at its end. When the tip is brought near a sample's surface, forces between the sample and the tip lead to the cantilever being deflected some amount. The amount of deflection is usually measured using a laser spot that is reflected from the top surface of the cantilever. This information is sent to a computer which generates a three-dimensional image of the sample. Depending on the sample being visualized, the AFM can operate in contact mode, where the sample moves back and forth under the cantilever, or tapping/non-contact mode, where the cantilever is vibrated across the sample surface.

Another microscope in the probe family is the scanning tunneling microscope (STM). The STM is unique in that its purpose is to image surfaces down to the atomic level. Its resolution of around 0.01 nm allows for both imaging and manipulation of individual atoms. In order to visualize a sample using STM, it must be electrically conductive, but the samples can be imaged in a variety of environments including high vacuum, ambient air, water and other liquids, as well as at a wide range of temperatures. The STM uses the concept of quantum tunneling to create images. Tunneling current is a quantum mechanical phenomenon that occurs when a conducting STM tip is brought near to the surface to be examined. When a voltage difference is applied between the tip and the surface, electrons can tunnel through the vacuum that exists between them. The result is a tunneling current that gets measured and displayed as a computer image.

We can also use x-rays to obtain information about the chemical composition of nanoscale materials. **X-ray spectroscopy** is a method that uses the characteristic x-rays produced by a material when it interacts with high energy electrons. **X-ray photoelectron spectroscopy** (XPS) is a technique that is often used when determining the chemical composition of thin films. XPS measures the elemental composition of the top 1-10 nm of a thin film. It must be conducted under ultrahigh vacuum.

NANOSPEAK – WITH ANDREW DOUCET (CHAPTER 7)

1. Where do you work and what is your job title?

    I'm going to college for a B.A.S in Electrical Engineering and am also a Process Technician for contamination free manufacturing at GlobalFoundries.

2. What is your educational background?

    Before working on my bachelor's degree, I had earned an A.A.S degree in Nanoscale Materials Science.

3. How did you get interested in nanotechnology/where did you hear about it?

    I became interested in Nanotechnology when I saw how big it was getting and I knew I wanted to get into a field that was going to have many jobs. I also love math and science and as of now, engineering as well!

4. What advice would you give to someone who might want to get into the field

    My advice to someone entering this field would be that while looking for jobs in the field of nanotechnology to not only consider the big companies, but to also look into some of the smaller more up-and-coming ones as well. There is a lot of room for moving up and being a part of something new and exciting.

CHAPTER 7 SUMMARY

- Top-down nanofabrication can be likened to sculpting figures from a larger block of material.
- The four basic steps in top-down nanofabrication are deposition, etching, material modification, and lithography.
- Deposition is the growth of thin layers from which nanostructures are formed. These layers can be grown onto a surface by a chemical reaction, physically applied to a material or grown via physical or chemical vapor deposition.
- Lithography refers to a patterning process that directs where material is deposited or removed. The most common form of lithography is photolithography, which uses a light-sensitive polymer called photoresist for surface patterning.
- Etching is an important process wherein materials are removed from specific areas of a substrate, using wet chemicals or dry plasmas.
- Material modification refers to any process that tailors the chemical, mechanical, optical, or electrical properties on a surface. Ion implantation is a common method to dope silicon in order to adjust its electrical character.
- Bottom-up nanofabrication can be likened to building a structure with individual building blocks. In nanofabrication, the building blocks are atoms, molecules, or nanoparticles that are synthesized and then spontaneously assembled into the desired structures.
- Because we cannot see nanostructures by eye, we use microscopes to visualize them. Some common microscopes are scanning electron, transmission electron, scanning probe, and scanning tunneling.
- X-rays can be used to obtain information about the chemical structure of nanomaterials via X-ray spectroscopy.

REFERENCES

1. M. Quirk, J. Serda, *Semiconductor Manufacturing Technology, 1ˢᵗ Ed.,* Prentice Hall, 2000.
2. H. Xiao, *Introduction to Semiconductor Manufacturing Technology,* Prentice Hall, 2000.
3. D. Vollath, *Nanomaterials: An Introduction to Synthesis, Properties, and Applications,* Wiley-VCH, 2013.
4. D. Murphy, *Fundamentals of Light Microscopy and Electronic Imaging, 2ⁿᵈ Ed.,* Wiley-Blackwell, 2012.

CHAPTER 8   BIOLOGICAL AND MEDICAL APPLICATIONS OF NANOSCALE MATERIALS

KEY TERMS

BIOMIMICRY	CANCER CELLS
LOTUS EFFECT	TUMORS
PLASTIC STEEL	TARGETED DRUG DELIVERY
GECKO TAPE	BIOAVAILABILITY
GLIOBLASTOMA	PHOSPHOLIPIDS
BIOSENSOR	CELL MEMBRANE
TRANSDUCER	LIPOSOME
BIOLOGICAL ELEMENT	BIOCOMPATIBLE
DETECTOR	SOLID LIPID NANOPARTICLES
ARRAY DETECTION TECHNOLOGY	

INTRODUCTION

Nanotechnology is a vast term that applies to many fields of science such as biology, physics, chemistry, and engineering. In chapter 8, the ways in which biology intersects with nanoscale materials will be examined. Biomimicry is a term that refers to any scientific attempts to emulate naturally occurring phenomena for the purposes of solving human problems. Nature takes advantage of nanostructures, for example in lotus leaves and gecko feet, which we try to synthetically copy.

Nanotechnology now plays a large role in the field of medicine, ranging from disease diagnostics to specialized disease treatment. Perhaps the greatest achievements have come in the form of targeted drug delivery, wherein nanomaterials are used to treat only specific areas of the body as opposed to much more invasive and systematic treatments of the past.

Some things to think about:
- What are some naturally occurring phenomena we have yet to be able to recreate in science?
- How are nanoscale materials helping to improve the field of disease diagnostics?
- What are the benefits and challenges associated with targeted drug delivery?

8.1   BIOMIMICRY

Biomimicry refers to any science by which we try to emulate naturally occurring phenomena to solve human problems. Nanotechnology is being used to mimic biology in attempts to reproduce the sophistication of nature in a laboratory setting. One example of nano-biomimicry is known as the **lotus effect**. The lotus plant is an aquatic species similar to a water lily. Its leaves demonstrate a special phenomenon known as superhydrophobicity. In superhydrophobic species, there is a very complex surface architecture that minimizes water adhesion and allows dirt to be picked up by the tight water droplets. Generally, water droplets will try to minimize the amount of surface tension experienced by forming a spherical shape. Of all geometric shapes, a sphere has the smallest surface area. Depending on the structure of the surface, the water droplet will experience some adhesion forces, known as wetting a surface. The degree of surface wetting may be measured using the contact angle where the drop meets the surface. The smaller the contact angle, the more the water drop has wet the surface. In the case of superhydrophobicity, a contact angle larger than 150° is generally attained.

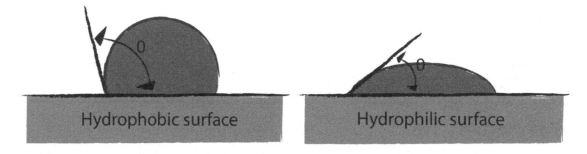

FIGURE 8.1

Figure 8.1 Contact angle measurements

Using nanotechnology, researchers are developing treatments for surfaces that mimic that of a lotus leaf. The surfaces would stay dry and be self-cleaning. The German company BASF is working on a lotus-effect aerosol spray that combines Teflon nanoparticles with hydrophobic polymers and waxes. As the spray dries, the material self-assembles into a thin coating on surfaces ranging from paper and masonry to leather and textiles. The goal is to make new products such as self-cleaning shoes.

Another example of nanotechnology mimicking biology is a new material known as **plastic steel**. Research out of the University of Michigan led to the formation of a composite plastic that is as strong as steel but is light and transparent. Plastic steel is modeled after a brick-and-mortar type of structure found in the iridescent lining of oysters, mussels, or abalone shells. It is also the same material that creates a pearl. Organisms, such as the abalone or oyster, layer an organic material with an inorganic material to create shells that are strong and lightweight. The structure has a nanoscale pattern that is similar to brick and mortar.

The layers of plastic steel are made of nanoscale-thick sheets of clay with a water-soluble polymer that acts as the mortar. The water-soluble polymer is very similar to white school glue. While the material is technically too brittle to be deemed plastic, this development could lead to lighter and stronger armor for soldiers or for new police vehicles. The composite is built in a custom machine that proceeds via a layer-by-layer assembly, just like a pearl grows in an oyster. There are approximately 300 layers of nanoscale clay and polymer glue that form a sheet of plastic steel as thick as a piece of plastic wrap.

Gecko tape is a material made by mimicking the ability of a gecko to scale walls and hang from ceilings. The gecko's secret method of climbing has only recently been unlocked. We now know that a gecko's feet have toe pads that are made up of about 500,000 fine hairs. These hairs, also called setae, are made of keratin, the same fibrous protein that makes up our hair and nails. Each hair has even smaller nanoscale projections protruding from their ends called spatulae. While scientists once thought that the gecko's climbing ability was the result of friction or suction, American biologist Robert Full at the University of California, Berkeley, found that the adhesion was a result of van der Waals forces between the spatulae and the surface. Andre Geim (the same physicist who was awarded a Nobel Prize for his work with graphene) then added that not only were van der Waals forces at play, but that capillary forces also contribute to a gecko's ability to adhere to a surface. Capillary force refers to attractive forces that are created by the surface tension of a molecular layer of absorbed water that form between two surfaces.

Andre Geim and others at the University of Manchester were able to synthesize a material that mimics the spatulae of a gecko which they called gecko tape. The process to make gecko tape involves several nanofabrication techniques, including the synthesis of a thin film substrate on a silicon wafer patterned via an aluminum mask created using electron beam lithography. Unlike conventional tape that we are most familiar with, gecko tape sticks when it slides across a surface in a given direction. Like a gecko's foot, the tape grips easily with sliding and then is very easily released from a surface. The tape may someday be used for medical applications, sporting goods, or climbing robots.

8.2   MEDICAL IMAGING AND DISEASE DIAGNOSIS

Nanotechnology is changing the way in which we can visualize biological materials for medical purposes.

Nanoparticles are being used for diagnostic imaging that can be used in the early detection of neurological diseases, cardiovascular disease, and cancer. The technology uses nanoparticles known as contrast nanoparticles. As previously mentioned, an MRI contrast agent is defined as a medium that helps to improve the visibility of structures during magnetic resonance imaging. A number of different nanoparticles can be used as contrast nanoparticles, including iron oxide and gold.

Contrast nanoparticles can be functionalized with biological targets so that they bind to specific areas of the body. To detect cardiovascular disease, for example, contrast nanoparticles can be functionalized to bind with proteins that are found on the surface of plaques that block arteries and can lead to heart attacks or strokes. When the functionalized contrast nanoparticles bind to the plaque proteins, they appear highly luminescent when visualized with a traditional MRI scan.

A similar technique may be used for imaging of cancer cells. Gold nanoparticles have been functionalized to target specific cancer cells, for example, a team at Stanford used them to help illuminate brain tumor cells. Approximately 3,000 people each year are diagnosed with a particularly aggressive form of brain tumor known as a **glioblastoma**. Once a patient is diagnosed with this form of brain tumor, the prognosis for the patient's survival is typically around three months. The tumors need to be surgically removed, but surgeons have a particularly difficult time with this surgery, as they need to remove all cancer cells without removing any normal brain cells. Glioblastomas have particularly rough edges with finger-like projections that grow into the healthy brain tissue. Patients with glioblastomas also tend to have micrometastases, or small tumor patches that come from cells migrating away from the main tumor. These micrometastases are generally invisible to a surgeon but can form into new larger tumors.

Functionalized gold particles were tested in mice with human cells implanted into their brains by intravenous injection. The gold nanoparticles bind only to tumor cells, and not healthy brain cells. The gold cores allow the particles to be easily visualized by MRI. These MRI images help to determine the tumors exact boundaries and give surgeons a much better picture of where to operate.

In addition to MRI imaging, other visualization methods have been tested in conjunction with nanotechnology. One group has used quantum dots to illuminate the inside of cells. The quantum dots can be functionalized to be attracted to specific proteins within whichever part of the cell a researcher wishes to examine. The benefits of using quantum dots, besides the ease at which their surfaces may be functionalized, also include their long life span. Other fluorescent dyes do not last very long, and do not allow for monitoring changes within a cell over a long period of time.

A research team at the National Institute of Allergy and Infectious Disease used this quantum dot technique to monitor cells infected with malaria. The quantum dots were functionalized to target a protein that forms part of a red blood cell's inner membrane structure. It was found that when the cell was infected with the malaria parasite, the structure of this particular protein changes. Being able to study these protein changes over time with quantum dot imaging allowed the group to study the progression of the disease in a cell infected with malaria.

Nanotechnology is also having a large impact in the field of biosensors. A **biosensor** is a device that responds to specific chemical species within biological samples. More specifically, biosensors integrate biological elements with a physiological chemical transducer to produce an electronic signal which is conveyed to a detector. The biological element is a component that is used to bind to a specific target molecule. It needs to be highly specific, for example one particular substrate that binds with only one enzyme. A **transducer** is simply a device that converts energy from one form to another form. When the enzyme and substrate bind, the physical change produces some amount of energy at that **biological element** site. The transducer would transform that energy into a measurable electrical output. The third component is the **detector**, which takes the signal from the transducer that gets passed to a microprocessor, amplifies and analyzes it. One biosensor you may be familiar with is a glucose meter, which determined the concentration of glucose or sugars in the blood.

Several different versions of biosensors have been made using nanoscale materials, for example biosensors that use nanowires, nanoshells, and nanotubes. These nanobiosensors have been used to detect a number of biological materials such as DNA, bacteria, viruses, and enzymes. A portable nanosensor that can quickly detect bacterial infections is being developed and has the potential to become commercially available in the near future. Currently, bacterial infections are diagnosed by swabbing a sample from a patient and having the sample grow in a laboratory, often over a period of a day or more.

One new nanosensor will detect the DNA of certain bacteria, and only take 15 minutes to an hour for a diagnosis. The sensor has hairpin-shaped strands of DNA attached to the surface of silver nanoparticles with a fluorescent molecule attached to the DNA (the strand of DNA that is complementary to the strand for the specific DNA being targeted). The silver quenches the glow of the fluorescent molecule. The DNA will stay folded in the hairpin shape until the targeted sequence links to it. When this happens, the DNA strand unfolds. This unfolding of the DNA strand results in the fluorescent molecule moving away from the silver nanoparticle, so it will glow. The glowing can be seen quickly using a fluorescent optical microscope.

A company in New York is working to commercialize disposable cartridges that contain the nanoparticle-DNA composites. A blood or urine sample can be tested by being placed directly into the cartridge. The cartridge can then be placed in a small, portable instrument capable of fluorescent imaging and analysis. If multiple strands of DNA were attached to the nanoparticles, the same cartridge could potentially screen for a variety of bacteria.

The nanopore immunosensor is a device used to detect the protein found in peanuts called Ara h1. Many people suffer severe peanut allergies. This particular biosensor uses nanoporous polycarbonate, a flexible polymer that can be shaped into many various forms. The pores of the nanoporous polycarbonate are coated with gold, and the antibody to the Ara h1 is anchored onto the gold. The peanut protein is detected when it binds to the anibody and changes the pore's electrical conductivity, since the pores are partially blocked when the antibody and protein bind together. This sensor is very specific to peanut proteins, and can detect trace amount of allergen, which is often ingested accidentally by peanut allergy sufferers.

Array detection technology refers to nanoscale arrays of pads that can be constructed onto a single chip using standard nanofabrication patterning techniques. As with other biosensors, various probe molecules such as DNA strands or specific antibodies can be attached to these pads. For example, the DNA can be tagged with quantum dots and fluoresce when the corresponding complementing strands links to it. In this way, specific pathogens can be detected if certain positions on the chip fluoresce. These arrays are often known as "lab on a chip." An example of this array detection technology is called "NanoChip®." A selection of probe DNA for various diseases is placed on the different chip pads. If the complementary strand binds to the probe DNA, the NanoChip will confirm presence of that specific disease or genetic defect.

8.3 DRUG DELIVERY

Often times, conventional medication is highly invasive throughout the entire human body. One example of this is current chemotherapies given to cancer patients. Cancer is the abnormal division and growth of cells that tend to proliferate in an uncontrolled and unregulated way and, in some cases, to metastasize. A **tumor** consists of a population of these rapidly dividing and growing **cancer cells** with mutations occurring quite rapidly. Tumors cannot grow beyond a certain size due to a lack of oxygen and other essential nutrients, which normally are supplied by the vascular system to all cells of our body. These tumor cells have the ability to induce angiogenesis by secreting various factors, thereby redirecting blood vessel growth towards them instead of other normal cells and tissues, thus denying these normal cells from getting their due share of oxygen and nutrients. Angiogenesis is a physiological process which allows for the formation of new blood vessels from pre-existing ones. It is a necessary process for growth and development as it allows for the delivery of nutrients and oxygen to our body cells. However, it is a necessary step also for the transition of dormant tumor cells to a malignant situation. By redirecting blood vessel growth towards them, the tumor cells can travel by blood and metastasize to other parts of the body. Eventually, the tumor can become so large that it crowds vital organs. These organs then cannot perform their necessary functions, and a patient may die as a result. Cancer, once detected, can be treated in a number of ways including surgery, radiation therapy, and chemotherapy.

Traditional chemotherapy drugs are delivered intravenously and are designed to kill any cells that divide rapidly. While cancer cells are indeed cells that grow at a rapid rate, humans also have cells that divide rapidly under normal circumstances, which include digestive tract cells, hair follicles, and bone marrow. As a result, both cancer cells and normal, healthy cells are attacked by the chemotherapy drugs. Patients tend to suffer many side effects such as hair loss, a suppressed immune system, and severe inflammation of the digestive tract lining.

In attempts to improve the current methods by which diseases are fought, scientists are using nanotechnology for what is known as targeted or "smart" drug delivery. The goal of targeted drug delivery is to send a concentration of drugs only to the localized areas where it is needed. This method should help to dramatically decrease negative side effects, and increase effectiveness of drugs for their intended purposes.

Nanotechnology has the potential to completely revolutionize the way in which drugs are delivered. Because we can so easily functionalize small nanoparticles, drugs can be delivered only to specific cells. By avoiding the current system which entails systemic delivery of drugs, this idea of targeted drug delivery can significantly reduce the amount of drug administered, the cost of therapy, and significant side effects experienced by many patients.

Bioavailability refers to the presence of a drug where it is needed in the body and where the drug can be most effective. The approach of targeted drug delivery is to improve bioavailability as much as possible. It is currently estimated that more than $65 billion dollars a year are wasted as a result of poor drug bioavailability. The combination of using nanoscale materials to both improve imaging and diagnostics as well as increase bioavailability should help to overwhelmingly improve effective in treating illnesses such as cancers.

8.3.1   LIPOSOMES

There are a variety of ways in which drugs are currently delivered, including intravenous delivery, topical delivery to the surface of the body, nose sprays, inhaled aerosols, eyedrops, oral pills, and injections, for example. One way to more effectively deliver drugs is to help get them through the membranes of cells. Cell membranes are made of amphiphilic molecules called **phospholipids**. An amphiphile is a molecule that has two parts: one that is hydrophilic (water-loving) and one that is hydrophobic (water-hating). On one end of the phospholipids that make up our cell membranes is called the head, which is hydrophilic. The other end contains two tails which are hydrophobic.

Our **cell membranes** are made of a bilayer of phospholipids, or two layers, which help separate the inside of the cell from its outside environment. The hydrophilic heads make up the outside layer of the membrane, while the hydrophobic tails are located in the inner layer of the membrane. This structure was evolved to help control the movement of substances into and out of cells.

The protective nature of the cell membrane also tends to inhibit therapeutic drugs from crossing to the inside of a cell. Nanotechnology is helping to find ways to ensure the correct drugs are able to enter cells. One way to deliver drugs into a cell is via a nanoparticle known as a **liposome**. A liposome is an artificial vesicle, or small pocket, also made of phospholipid bilayers. These liposomes can be used as a nanoscale vehicle for drug delivery. By adding the correct concentration of phospholipid molecules to an aqueous solution containing a specific therapeutic agent, the molecules will self-assemble into a spherical liposome and encapsulate the drugs in the process.

The result of this process is a small drug-carrying liposome which has the ability to fuse with cells. Once the liposome has fused with a cell, it can release its contents to the inside of a cell.

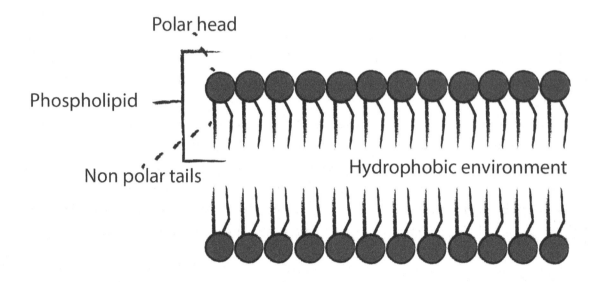

FIGURE 8.2

Figure 8.2 Phospholipids

Cisplatin is a well-known chemotherapy drug. The cisplatin molecule contains an atom of platinum at its center which can bind to and alter DNA in rapidly growing cells, causing them to burst and die. The drug has been used to treat a variety of cancers but has been particularly effective against testicular cancer. Despite its success as an anticancer agent, cisplatin causes many adverse side effects that result from systemic dosing, including kidney damage, nerve damage, nausea, vomiting, and hearing loss.

Recently, it has been found that cisplatin can be encapsulated in liposomes. This liposomal Cisplatin complex is often referred to as lipoplatin. Studies have shown that lipoplatin is significantly less toxic than cisplatin, and that it is still highly effective in reducing the incidence of tumors in patients. So far, lipoplatin treatment is showing a great deal of promise because of its negligible toxicity and high effectiveness.

8.3.2   METALLIC NANOPARTICLES

In addition to liposomes, there are a series of metallic nanoparticles that have great promise in the field of targeted drug delivery. One such nanoparticle system is iron oxide. Iron oxide nanoparticles, as previously discussed, are magnetic. These magnetic iron oxide nanoparticles, or MIONs, can be guided to a targeted location using an external magnetic field. To keep our immune systems from attacking foreign substances such as MIONs and allow them into the cell membrane, they can be coated in a biocompatible material such as the polymer polyethylene glycol (PEG), or other biomolecules such as proteins. A molecule which is biocompatible refers to any molecule which goes undetected by the immune system and may freely flow through the bloodstream without being attacked. MIONs can be modified with therapeutic drugs and the drugs released at a required location after being guided by magnets.

The MagForce Iron Oxide Nanoparticle Cancer Therapy is a treatment which has been developed in the European Union. Coated MIONs are injected directly into a brain tumor and the patient is exposed to an alternating magnetic field. The magnetic field pulses help the MION system get absorbed into tumor cells. Once inside tumor cells, the particles cause the cells to die and burst open. The dead cells are then naturally discharged from the body. This treatment is minimally invasive and was approved for cancer treatment in 2011.

Gold nanoparticle systems have also been explored as an option for a drug delivery vehicle. There are a wide range of studies reporting the use of gold nanoparticles as drug delivery vehicles. Tumor necrosis factor-alpha (TNF-α) is a drug with excellent anticancer efficacy, but is incredibly toxic and causes a severe systemic inflammatory response, limiting its use as a chemotherapy agent. In initial studies of TNF-α in the 1980's, it was found that the dose at which half the patients died was actually lower than the effective systemic dose.

Research has now shown that functionalizing gold nanoparticles with TNF-α allows the nanoparticles to specifically target tumor cells in a small enough concentration that there are fewer harmful side effects than in system dosage, but high enough concentration to cause cancer cells to be killed. Attaching TNF-α to gold nanoparticles actually makes the drug less toxic. The gold nanoparticles are also coated in a biocompatible polymer, as with other nanoparticle drug delivery systems, so that they can travel though the body undetected by the immune system.

Gold nanoparticles may also be used to "cook" cancer cells. Research has shown that gold nanoparticles may be functionalized with proteins to specifically enter cancer cells. After collecting inside of the malignant cells, the area can be exposed to near-infrared light, between 800 and 2,500 nm. Gold nanoparticles absorb light in the near infrared very efficiently. Once the light is absorbed, heat is generated. This heat causes the cancer cells to be damaged and die, literally being "cooked" from the inside out.

8.3.3 POLYMER AND SOLID LIPID NANOPARTICLES

Polymeric nanoparticles made from biocompatible polymers such as polyethylene glycol are yet another innovative method of introducing medication to specific sites in the body. There are several advantages of using these particles made of polymers. For example, they highly effective at targeting specific cells, and can enable careful release of any drugs that are incorporated in the system. This is an advantage over systems such as liposomes, where therapeutic agents often can diffuse out of the vehicle before they reach the desired area of the body.

However, the polymers have been shown to be toxic to healthy cells after they are internalized. Also, it can be very difficult to produce a large scale amount of polymeric nanoparticles. For these reasons, this carrier system may not have future promise in the pharmaceutical industry.

Solid nanoparticles made up of lipid molecules are a potential improvement over polymeric nanoparticles. **Solid lipid nanoparticles** (SLNs) are able to carry drugs in a vehicle that is composed of physiological lipids dispersed in water. Again, a lipid is a naturally occurring molecule such as fats, waxes, and vitamins, which can be completely hydrophobic or can be amphiphilic. When referring to SLNs, they can be made of a solid triglyceride, diglyceride, fatty acid, steroid, or wax core surrounded by water-stabilizing molecules. The therapeutic agents are carried inside of the lipid core.

The obvious advantage to using SLNs for drug delivery is that they pose no potential toxicity problems that other non-naturally occurring nanoparticle systems may. In addition to being non-toxic, there is better controlled release of the therapeutic agent and it can be delivered over a long period of time. The surfaces of SLNs are also easily modified to target specific cells. They are more stable than liposomes, are biodegradable, and can be stored for a long period of time.

NANOSPEAK – WITH ROBERT BROWN (CHAPTER 8)

1. Where do you work and what is your title?

    I currently am working for Atotech USA. Atotech is an international company that specializes in electrochemical plating of metals. I work for the semiconductor branch of this company wherein we plate metals for items as large as car parts to as small as the M1 layer of a microchip.

2. What is your educational background?

    I have a Bachelor of Arts in Mathematics, a Master's of Science in Secondary Education, an Associates of Applied Science in Nanoscale Materials Technology, and plan to begin work on my Masters in Nanoscale Engineering.

3. How did you get interested in nanotechnology/where did you hear about it?

    After a few years of teaching math, I found that wasn't the career for me. Given my background, I knew I wanted to do something math or science related for a career, I just didn't know what. I had a few roommates that were scientists who worked for GE and that got me started to think about what careers I might like. During this time, I learned that semiconductor manufacturing was expanding rapidly. It was clear to me that nanotechnolgy would be a major presence in the area where I wanted to stay and make a home. After taking a visit to the Albany College of Nanoscale Science and Engineering, I was hooked!

4. What advice would you give to someone who might want to get into the field?

Make as many contacts as you can. Something as simple as a friendly casual conversation can lead to an internship. That's what happened for me. In this industry, you work closely with people on a daily basis. Making strong working relationships with your colleagues, professors, fellow students, etc. will help you succeed and stand out.

CHAPTER 8 SUMMARY:

- Biomimicry refers to any science by which we try to emulate naturally occurring phenomena to solve human problems.
- Some examples of naturally occurring nanostructures we have tried to recreate in a lab are gecko feet, lotus leaves, and abalone shells.
- Nanomaterials can be used for disease diagnostics, for example, detection of specific neurological diseases, cardiovascular diseases and cancer.
- Nanoparticles can be used for medical imaging such as Magnetic Resonance Imaging.
- Some biosensors employ nanoparticles. Biosensors integrate biological elements with a chemical transducer to produce an electrical signal as in a glucose sensor.
- Nanotechnology is changing the face of drug delivery. Often conventional therapies are highly invasive. Use of nanoparticles can help to deliver therapeutic agents only to necessary areas. Some nanoparticles employed for such an application are liposomes, iron oxide, gold, polymers, and solid lipid nanoparticles.

REFERENCES

1. D. Goodsell, *BioNanotechNology: Lessons from Nature*, Wiley-Liss, 2004.
2. M. Amiji, *Nanotechnology for Cancer Therapy*, CRC Press, 2006.
3. R. Booker and E. Boysen, *Nanotechnology for Dummies, 1st Ed.*, For Dummies, 2005.

CHAPTER 9   NANOTECHNOLOGY AND ENERGY

KEY TERMS

SOLAR POWER	WET CELL BATTERY
PHOTOVOLTAIC CELL	DRY CELL BATTERY
PHOTOELECTRIC EFFECT	THERMAL CELL BATTERY
P-N JUNCTION	PRIMARY BATTERY
SOLAR PANEL	SECONDARY BATTERY
CIGS	POWER DENSITY
FUEL CELL	CAPACITOR
CATHODE	ULTRACAPACITOR
ANODE	GENERATOR
ELECTROLYTE	NANOGENERATOR
ELECTROLYSIS	PIEZOELECTRIC
BATTERY	TRIBOELECTRIC
ELECTROCHEMICAL CELL	PYROELECTRIC

INTRODUCTION

One of the largest issues the human race currently needs to address is the need for new and sustainable energy sources. Currently, oil, coal, and natural gas supply the vast majority of the world's energy supply. Coal is generally used to make electricity, natural gas for heating, and oil to power our machines. These energy sources are finite, however, and as they become scarcer, the prices will continue to rise. If we continue to use these resources at our current rate, there would be only enough oil to last approximately 40 years, enough coal for 130 years, and enough natural gas for 60 additional years. The era of non-renewable energy sources is coming to an end, and nanotechnology may be able to help address this enormous challenge of finding new ways to generate sustainable energy.

In Chapter 9, some ways in which nanoscale materials are being explored for potential solutions to our energy crisis will be discussed. Nanomaterials can be fabricated into thin solar panels, parts for fuel cells, components of batteries, capacitors, and even nanogenerators.

SOME THINGS TO THINK ABOUT:

- What are some of the issues associated with commercial solar devices?
- How are fuel cells similar to batteries? How are they different?
- How is nanotechnology helping solve some major roadblocks in the development of commercially available fuel cells?
- What different types of batteries exist? How might they be used?

9.1   SOLAR ENERGY

The sun produces light and heat which can be harnessed as solar energy. The ability to take advantage of solar energy in an affordable manner could provide us with a clean technology with excellent long-term benefits.

Every day, the earth receives a large amount of incoming solar radiation. This electromagnetic radiation is generally in the visible light region of the spectrum, with some in the infrared and some in the ultraviolet regions as well. About 30% of this radiation gets reflected back into space in the upper atmosphere. The rest of the radiation enters our atmosphere and is absorbed by clouds, the oceans and land. The amount of solar energy that reaches the Earth's surface is immense. More solar energy reaches the Earth's surface in one year to equal about twice as much energy as can be obtained from oil, coal, and natural gas combined.

Solar power refers to the conversion on the entering solar energy into electricity. This can be accomplished using devices known as **photovoltaic** (PV) **cells**. PV cells, or solar cells work by converting light into an electric current via the photoelectric effect. In the photoelectric effect, matter absorbs energy in the form of electromagnetic radiation and emits electrons which can act as charge carriers for an electric current.

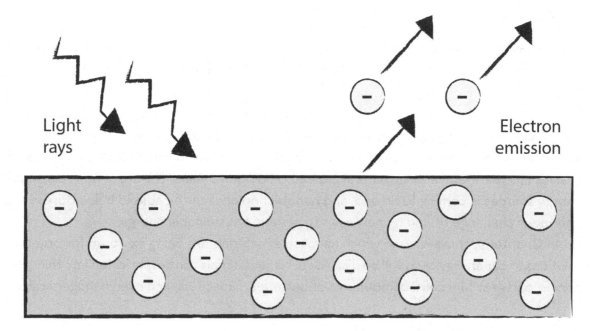

Light rays

Electron emission

FIGURE 9.1

Figure 9.1 The photoelectric effect

A solar cell is primarily made of silicon. The bottom part of the silicon is doped with a trivalent impurity to accomplish a p-type semiconductor, and the top is doped with a pentavalent impurity to form an n-type semiconductor (Section 3.4.3). As sunlight hits the silicon atoms in the solar cell, their energy is transferred to electrons that are knocked loose via the **photoelectric effect**. Once the electrons are free, they need to be able to create an electric current. This is accomplished by generating an electrical imbalance within the cell, causing the electrons to flow in one direction. The electrical imbalance results from having a p-type semiconductor adjacent to an n-type semiconductor and forming a **p-n junction**. At the p-n junction, the extra valence electrons from the n-type semiconductor can jump over to fill in the holes of the p-type silicon. This means that there is a small region at the p-n junction where the n-type silicon becomes positively charged from missing electrons, and the p-type silicon is negatively charged from gaining the extra electrons.

This creates a directional electric field across the solar cell. As the arriving sunlight frees electrons from the silicon atom, the electric field created as a result of the p-n junction drives the free electrons along in an orderly manner, providing an electric current that can be used for various applications. Because each PV cell only produces a small amount of voltage, multiple cells are combined by manufacturers into **solar panels** which are capable of producing larger, more useful amounts of electricity.

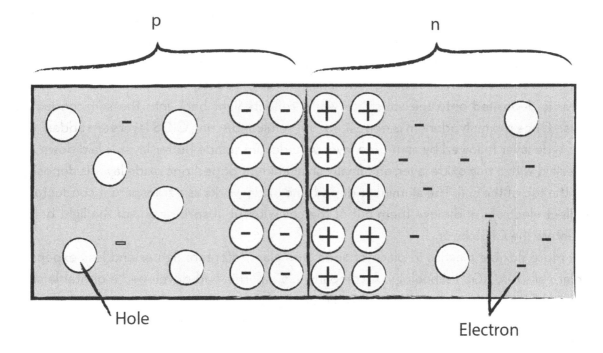

FIGURE 9.2

Figure 9.2 P-N Junction

While using solar energy to create electricity via solar panels is one of the cleanest possible methods of energy production with no harmful byproducts, it is not without its problems. One major concern with current solar panel technology is the cost to manufacture, which makes solar panels incredibly expensive to purchase, install, and maintain. Silicon and the surrounding glass in solar panels are very fragile, and consistent maintenance and replacement of cells are generally required. There is a large amount of high quality silicon needed, which drives up costs. It also requires a great deal of energy to produce solar panels. Often, more energy is required to produce solar cells than they are able to save over their lifetime.

Nanoscale materials are being used to create a new generation of solar cells that can generate electricity at lower cost using a much smaller amount of material. Nanoscale materials can also help to make solar cells more efficient, and a more viable source of energy production.

There are several examples of nanotechnology in photovoltaics. One way to make the manufacture of solar panels less expensive is to reduce the amount of bulk silicon needed. As will be discussed in Chapter 11, the production of bulk electronics grade silicon from silicon dioxide is a costly and lengthy process. The process of manufacturing electronics grade silicon into functioning devices is additionally complex and expensive. The use of nanoscale thin films for solar cells requires much less energy and much less material to produce.

Thin film solar cells are being created using a semiconductor material known as **CIGS**, or copper indium gallium diselenide. The CIGS material absorbs sunlight very strongly, and therefore only a very thin film is required for producing PV cells. The most common structure for CIGS solar cells involves using soda lime glass as the substrate. A thin film molybdenum layer is deposited onto the substrate which reflects light back into the semiconducting layer. Once the molybdenum is deposited, the semiconducting CIGS layers are added; first a p-type layer followed by an n-type layer. A cadmium sulfide buffer layer is laid down and overlaid with a zinc oxide layer, and finally an aluminum-doped zinc oxide layer is deposited on the top of the cell. The aluminum-doped zinc oxide works as a transparent conductor to collect electrons and move them out of the cell without absorbing any of the light before it enters the CIGS layer.

More flexible plastics in place of soda lime glass to create lighter and less expensive solar cells for CIGS technology. A company in California had developed a printable solar cell "ink" that can be printed onto flexible plastic substrates.

Another technology that may replace current silicon solar panels is organic solar cell technology. Organic solar cells don't use semiconductor layers at all, but instead, use organic, carbon-containing molecules to act as the p-type and n-type layers. They can be manufactured on inexpensive plastic for a lightweight and flexible alternative to current silicon solar cell technology. The main disadvantages associated with organic solar cells are their low efficiency of converting sunlight into electricity, and potential low stability of the organic molecules.

9.2 FUEL CELLS

A **fuel cell** is a device that, like a solar cell, creates electricity. Fuel cells generate electricity via a chemical reaction between oxygen and an oxidizing fuel source. Most commonly, hydrogen is used as the fuel, but organic alcohols like methanol can also be used. In a hydrogen fuel cell, hydrogen is combined with oxygen to produce electricity with byproducts of heat and water.

In general, a fuel cell is constructed with a design similar to that of a battery. Fuel cells contain a **cathode** (positively charged electrode), an **anode** (negatively charged electrode), and an **electrolyte** that allows the charge to move between the anode and the cathode. Unlike batteries, however, fuel cells require a constant source of fuel and oxygen to run. The fuel cell must also contain a catalyst that breaks apart hydrogen and oxygen molecules into ions with the production of free electrons. Hydrogen and oxygen pass through a porous carbon electrode into the aqueous electrolyte solution.

The chemical reaction in a hydrogen fuel cell involves hydrogen being delivered to the anode where it is oxidized by hydroxide ion (OH-), which is produced by feeding oxygen into the cathode where it becomes reduced. The net reaction is the formation of water and energy from hydrogen and oxygen. The anode and cathode are made of porous compressed carbon that contains a small amount of catalyst particles, such as platinum or silver. As long as hydrogen and oxygen are continuously fed into the cell, hydrogen and oxygen can continue to be converted into energy.

Since fuel cells are highly efficient, provide a continuous source of energy, and work without creating any pollution, they make an attractive candidate for replacing oil-powered vehicles. Fuel cells are also much lighter in weight than the batteries that are required to power electric cars. Despite these many advantages, however, hydrogen fuel cells and the cars they might power are not without their drawbacks.

One major issue associated with hydrogen fuel cells is the production of the hydrogen fuel that is required. Hydrogen gas is produced by a process known as steam reforming, or "cracking" methane (natural gas). Methane is a fossil fuel, which means it is not a sustainable starting material. There are two main steps in the steam reforming of methane. First, gaseous water, or steam, at 700-1100°C reacts with methane to generate carbon monoxide and hydrogen gas. In a second step, the carbon monoxide is further reacted with water at 130 °C to produce more hydrogen and carbon dioxide. While the process produces the valuable hydrogen gas, carbon dioxide is also produced. Carbon dioxide is categorized as a greenhouse gas, or gas that contributes to dangerous warming of the earth's surface.

Hydrogen can also be generated from the **electrolysis** of water according to the following reaction:

$$2\,H_2O \quad O_2 + 2\,H_2$$

The electrical energy required for water electrolysis generally comes from a fossil fuel source, which somewhat negates the environmental benefits of a clean fuel source such as hydrogen.

In addition to the production of hydrogen as a fuel source, the storage of hydrogen gas also presents a problem when designing vehicles that run on hydrogen fuel cells. Hydrogen gas has very poor energy density by volume, meaning it requires very large tanks to store. One gram of hydrogen occupies about 11 liters of space at atmospheric pressure. In order to be stored in a car, for example, the hydrogen must be intensely pressurized and stored in a special vessel to prevent explosions. If the hydrogen is compressed to a liquid, it can only be stored under exceptionally cold temperatures, which is also not very practical for everyday use.

The use of nanotechnology may help to solve some of these fundamental hydrogen production and storage issues. There has been a push to develop sustainable methods for hydrogen production, for example, using solar energy. Several years ago it was shown that titanium dioxide nanotubes are very effective photocatalysts, and can be employed in the conversion of water into hydrogen and oxygen using the power of sunlight.

It has more recently been shown that gold nanoparticles may be able to very efficiently produce hydrogen from water using solar energy to activate the reaction. Using very small gold nanoparticles and solar energy, a group at Stony Brook University was able to generate hydrogen gas from water nearly 35 times more efficiently than with other metal catalyst.

In the area of hydrogen storage, nanomaterials may provide a solid surface that could store hydrogen at very high densities. The key to this research is to synthesize a material that can both absorb hydrogen at high densities and also release it efficiently. A group at Lawrence Berkley National Laboratories developed a composite material made of magnesium nanoparticles in a flexible organic polymer that can absorb hydrogen gas, store it safely at high densities, and then release it rapidly when needed. Additionally, it is known that hydrogen bonds well with carbon. Several studies have been conducted on graphene and it seems that the two-dimensional carbon sheets also have potential as a solid state hydrogen storage device. If the graphene sheets are chemically functionalized, they may have enhanced hydrogen absorption properties at standard temperature and pressure.

9.3   BATTERIES AND CAPACITORS

While generating energy is an important aspect of sustainability, energy also needs to be stored efficiently. The most common ways in which energy is stored is via batteries and capacitors.

The generic term '**battery**' refers to any device that converts chemical energy directly to electrical energy. Batteries, which were first invented in the 1800s by Italian physicist Alessandro Volta, are made up of one or more electrochemical cells, which are devices powered by oxidation-reduction reactions. Oxidation-reduction reactions are essentially reactions that involve the transfer of electrons from one chemical species to another. The term oxidation means electrons are lost from a species, and the term reduction means those electrons were gained. A battery simply takes advantage of the electric current produced by moving electrons in oxidation-reduction reactions.

Each electrochemical cell consists of two separated half-cells. One half cell contains an electrode to which negative charges move (anode) and the other has an electrode to which positive ions move (cathode). There is an electrolyte in the half-cells which facilitates travel of charge, and while the half-cell electrodes never physically touch, they are connected by a material which helps charge flow between them. This electrolyte material can be in a liquid form, which results in what is known as a **wet cell battery**, or there are **dry cell batteries** wherein the electrolyte is immobilized in a paste form. In dry cells, there is only enough moisture to enable charge to flow. A third type of electrolyte is made up of molten salt, for example liquid sodium salt. These batteries must operate at high temperatures, and are often referred to **as thermal cell batteries**.

Batteries can be categorized into two general types: primary and secondary. **Primary batteries** are also known as disposable batteries. They are intended to be used once and then discarded and can be used immediately upon assembly. **Secondary batteries** are also known as rechargeable batteries. They can be recharged by applying an electric current to them which reverses the chemical reaction that causes electrons to move. Among the first rechargeable batteries used were the lead storage batteries in automobiles, which have been in use since 1915. Lead storage batteries can function at a wide range of temperatures and under the constant shock of rough roads. Other common secondary batteries use chemistry based on metals such as nickel and lithium. These are the batteries that power devices such as laptops and other electronics, as well as industrial equipment and battery-electric automobiles. There is potential to improve the performance of secondary, rechargeable batteries using nanotechnology.

There are several safety concerns surrounding lithium ion batteries due to their size and potential for leakage. The batteries that are large enough to be used in hybrid automobiles, for example, carry extremely high voltage. Most hybrid vehicles contain two large lithium ion batteries that are connected in series to make one larger battery. Each one generates about 110 volts, or 220 total volts. This is equivalent to the amount of electricity needed to power an entire home. According to Ohm's Law, voltage is equal to current multiplied by resistance (**Equation 3.7**).

Lithium ion batteries carry high voltage and low resistance, leading to dangerously high amperage. Were you to hit something on the road and wires get uncovered, if you touch the wire the amperage can cause your heart to stop.

Nanoscale materials can help to improve the power density of rechargeable batteries. **Power density** is the amount of power a battery outputs per unit volume or weight of the battery. We are most interested in designing batteries that can store a larger amount of power in lighter packaging that takes as little space as possible. Lithium ion batteries, for example, have power density and longer lifetime (about two times higher) than nickel-based batteries, which is why they have been so popular in portable electronic devices. Since nanoscale materials have such an increased surface area, they can allow charge to flow more freely at the anode and cathode. This will increase the battery's storage capacity and will result in smaller batteries, and shorter amounts of time needed for recharging. Researchers are investigating carbon-coated silicon nanowires, carbon nanotubes, and metal oxide nanoparticles such as vanadium and manganese oxide as possible materials for batteries.

Liquid electrolytes can cause battery cells to rupture, and leaking electrolyte materials can combust upon overheating. To prevent this from happening, safety measures have been put into place to further separate cells, which increases the size and weight of batteries, but also add to the cost of manufacture. Nanotechnology can also be used to improve battery electrolyte materials. Again, the large surface area of nanoscale materials can enhance the conductivity of electrolyte ceramics and pastes. Aluminum oxide, silicon dioxide, and zirconium dioxide nanoparticles have been incorporated into polymer gels as new electrolyte materials that help to enhance safety, conductivity and storage capacity, and eliminate the chance of cell rupture.

Energy can also be stored using capacitors. **Capacitors** hold charge using at least two electrical conductors that are separated by an insulator, which is also known as a dielectric. When a voltage is applied to the conducting materials, an electric field develops across the dielectric material, causing charge to be collected. Capacitors can be charged much more quickly than batteries, but have a significantly lower power density.

Research is being conducted to make what are known as ultracapacitors using nanoscale materials. **Ultracapacitors** are a new type of capacitor that, like batteries, use electrochemistry as opposed to the conventional solid dielectric material used in conventional capacitors. By using carbon nanotubes to increase the surface area of the electrode materials, ultracapacitors can store a significantly larger number of electrons than their conventional counterparts. Current ultracapacitor technology has a far lower power density than lithium ion batteries, but ongoing research shows promise to develop new ultracapacitors that have power densities on the same order of magnitude as secondary batteries, with a much longer lifespan.

9.4   NANOGENERATORS

A **generator** is a device that converts mechanical energy into electrical energy. The mechanical energy for generators can come from wind turbines, steam turbines, or water falling through a turbine, for example. Nearly all of the power for our electric power grid comes from some form of generator. The current technology requires a constant supply of energy. Steam turbines, which are very commonly used to generate electricity, burn fossil fuels such as coal or oil to produce electricity, and therefore are not a sustainable source of electrical energy.

Nanogenerators are devices that also convert mechanical energy into electricity, but are self-powered and self-sustaining. While they would not currently provide sufficient energy for a power grid, they may be able to help replace secondary batteries for devices such as personal electronics.

There are three main types of nanogenerators:

- Piezoelectric
- Triboelectric
- Pyroelectric

The piezoelectric and triboelectric nanogenerators run on mechanical energy, while the pyroelectric nanogenerator uses thermal energy to generate electricity. Professor Zhong Lin Wang of the Georgia Institute of Technology has pioneered research in the area of nanogenerators and has demonstrated the use of all three types to power small devices.

Piezeoelectric materials generate an electric potential in response to an applied mechanical stress. Arrays of zinc oxide nanowires have been shown to generate electrical energy in response to the stress applied from an atomic force microscope probe running over their surface. There is a great deal of potential for these types of self-powered generators. For example, they could be integrated into clothing and generate electrical energy from the vibrations produced that result from our movements.

The **triboelectric** effect is a contact electrification that results when materials are rubbed together and become electrically charged. Many of us have conducted the experiment in which glass is rubbed with fur or rubbing a balloon on our hair which builds up triboelectricity. Most static electricity is triboelectric. Triboelectric nanogenerators take advantage of the electrical charge that results from materials being rubbed together. The triboelectric nanogenerator has two sheets stacked upon each other with electrode layers on the back. The two sheets are covered on the inside with two materials that have a difference in ability to attract electrons, just like hair and a latex balloon. There is a small gap between the sheets that allows them to rub together. When a small mechanical agitation is applied to the sheets to press them together, the inner surfaces rub and enable the useful transfer of electrons.

Finally, the **pyroelectric** nanogenerator works by the pyroelectric effect wherein materials are heated or cooled and generate electricity. The change in temperature causes the atoms in pyroelectric materials to change position. This then leads to a slight deformation of crystal structure. When the crystal is changed, the result is a temporary voltage across the crystal. Pyroelectric nanogenerators have already been used to partially charge lithium ion batteries by using the energy from temperature fluctuations in the environment.

NANOSPEAK – WITH DEREK PALMERI (CHAPTER 9)

1. Where do you work and what is your title?

I work as a technician at the College of Nanoscale Science and Engineering, for the Research Foundation of New York.

2. What is your educational background?

   My current educational background is an Associate in Applied Science in Nanoscale Materials Technology. I have plans to continue working towards a Bachelor's degree.

3. How did you get interested in nanotechnology/where did you hear about it?

   I got interested in nanotechnology when I was a culinary student at my local community college, and was not looking forward to working in that field and following in the footsteps of the same thing people have been doing for millions of years, which is cooking. I wanted to, in a way, leave more of a footprint of my existence on earth by becoming a part of the nanotechnology workforce, and playing a part in the evolution of technology by getting the education I needed to enter working in the field of nanotechnology

4. What advice would you give to someone who might want to get into the field?

   I would advise anyone who is interested in nanotechnology to research the history of how, where and what nanotechnology has begun from and transformed into, in such a brief amount of time, relative to how long humans have been on earth, and understand the reality of Moore's Law continuing to prove itself true to our technology. It is also very interesting to see what other industries nanotechnology is finding its way into, such as for military equipment for devices and self-healing gas canisters for tanks which can take an armor piercing bullet then immediately patch itself up to avoid leakage. In the medical field, targeted drug delivery is currently a very large focus, as well as developing new types of sensors that can sense cancers and tumors at an earlier stage than current technology can. The most important aspect of nanotechnology is to realize that by harnessing the potential of this field of science, we can create new materials out of the current finite types of materials we have access to, to theoretically create infinite types of materials to behave and exhibited desired properties based on the needs of the creator and consumer. It is the unlimited potential of nanotechnology which motivates me to be a part of this industry and become involved in changing the world.

CHAPTER 9 SUMMARY

- We need to explore sustainable forms of energy.

- The sun produces heat which can be harnessed as solar energy. Solar power is the conversion of solar energy into electricity. This is accomplished using photovoltaic cells. A group of photovoltaic cells is called a solar cell.
- Currently, solar panels are primarily made of silicon. The cells are expensive to manufacture, require maintenance, and a good deal of energy is wasted in their production.
- Nanoscale materials are being used to create new generations of solar cells that are thin, lightweight, and energy efficient.
- A fuel cell creates electricity via a reaction between oxygen and hydrogen. The process by which hydrogen gas is obtained is inefficient, and hydrogen can be difficult to store.
- Nanoparticles are being explored as a method by which hydrogen gas can be generated from water and stored.
- Batteries are a vehicle for energy storage. Batteries create energy via chemical reactions. Primary batteries are disposable, while secondary batteries are rechargeable. Nanomaterials may help to improve the amount of power a battery can produce and enhance their safety.
- Capacitors are also devices that hold charge. Ultracapacitors are new devices that use carbon nanomaterials to store energy.
- Nanogenerators are devices that use nanomaterials to convert mechanical energy into electrical energy.
- The three main types of nanogenerators are: piezoelectric, triboelectric, and pyroelectric.

REFERENCES

1. G. Masters, *Renewable and Efficient Electric Power Systems*, Wiley-IEEE Press, 2004.
2. S. Kodigala, *Thin Film Solar Cells From Earth Abundant Materials*, Elsevier, 2013.
3. S. Xu, Y. Qin, Y. Wei, R. Yang, Z. Wang, "Self-Powered Nanowire Devices," *Nature Nanotechnology*, **5,** 2010, p. 366

CHAPTER 10   ENVIRONMENT

KEY TERMS

CARBON CYCLE   CATALYST
GREENHOUSE GAS   WATER POLLUTION
GREENHOUSE EFFECT   PATHOGENS
INTEGRATED GASIFICATION COMBINED   AEROGEL
   CYCLE TECHNOLOGY   POINT-OF-USE
MEMBRANE   DESALINIZATION
PHOTOSYNTHESIS   ION CONCENTRATION POLARIZATION

INTRODUCTION

Many forms of pollution exist on our planet, nanoscale materials might be able to help clean our environment. Carbon dioxide is considered a greenhouse gas since it is able to absorb infrared frequencies of electromagnetic radiation and radiate them back to earth. Abnormal amounts of carbon dioxide added to the atmosphere by human activity contribute to climate change, which can have very dramatic effects on the earth. Nanotechnology may help to capture carbon dioxide and reduce its warming effects.

In addition to air pollution, water pollution is a serious concern and significant global problem. It has been suggested that water pollution is one of the leading causes of death and disease in the world. The ways in which nanotechnology can be used to clean contaminated water will be explored.

Some things to think about:

- What are the ways in which the human population contributes carbon dioxide to the atmosphere?
- What makes a gas classified as a "greenhouse" gas?
- What are some of your own ideas as to how nanotechnology can be used to clean contaminated water?

10.1   CLEANER AIR

Carbon dioxide (CO_2) is a naturally occurring compound made up of a carbon atom covalently bonded to two oxygen atoms. In our atmosphere it exists in the gas phase, and also exists as a gas at standard temperature and pressure. Carbon dioxide is an important part of the **carbon cycle**, wherein plants, algae and certain bacteria convert the gas into valuable life-sustaining sugars, with oxygen gas produced as a byproduct.

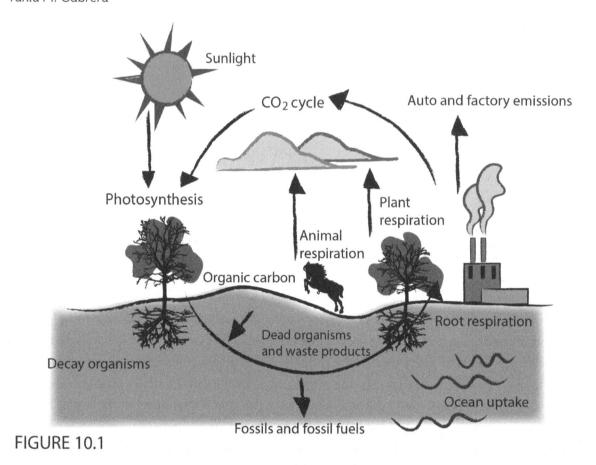

FIGURE 10.1

Figure 10.1 The carbon cycle

Carbon dioxide gas can be generated in a number of ways. For example, humans and land animals exhale CO_2 during the breathing process. CO_2 is what makes bubbles in beer due to sugars being converted to alcohol and carbon dioxide during their fermentation process. CO_2 is naturally emitted from the earth's crust via volcanic eruptions, from geysers and hot springs.

Humans have introduced a large amount of carbon dioxide into the environment in artifical ways, such as by burning coal for energy or combustion of gasoline and other hydrocarbons in automobiles. While maintaining certain CO_2 levels are important for life on earth, the continued burning of coal and oil has rapidly increased the concentration of carbon dioxide in our atmosphere. Carbon dioxide is classified as a **greenhouse gas**. This phrase dates back to the 1890s when Swedish chemist Svante Arrhenius published calculations showing that carbon dioxide gas in the atmosphere acts like the glass in a greenhouse, which is meant to trap in warm air and protect plants from cold weather. Arrhenius showed that the temperature of the earth's surface is very sensitive to the amount of carbon dioxide in the atmosphere, which led to the currently used term, **greenhouse effect**.

The main primary gases in our atmosphere are oxygen and nitrogen. They are transparent to visible light coming in from the sun. Remember from Chapter 9 that when sunlight reaches the surface of the earth, it is absorbed and converted into heat. Heat is simply kinetic energy which can be transferred to atoms on the earth's surface and cause them to increase in vibrational energy. The energy of the atoms vibrating is radiated in the infrared region of the electromagnetic spectrum. Both oxygen and nitrogen do not absorb infrared frequencies of radiation, so if our atmosphere contained only these gases, the warmth from infrared radiation would escape back into space.

Despite oxygen and nitrogen being the main components of our atmosphere, it does contain other gases, such as carbon dioxide. Unlike oxygen and nitrogen, CO_2 does absorb infrared frequencies and radiates them back toward the earth. This absorption helps to warm the atmosphere and the earth's surface, which is known as the greenhouse effect. While the exact relationship between concentration of CO_2 in the atmosphere and temperature is not completely clear, what is understood is that since 1980, the temperature of the earth's surface has risen about 0.8 °C, and scientists believe it is due to the increased levels of carbon dioxide added to the atmosphere by human activity and the widespread use of fossil fuels.

While 0.8 °C does not seem like a large number, subtle changes in global temperature result in dramatic consequences. The effects of increasing global temperatures could cause sea levels to rise, affect precipitation and extreme weather conditions, and result in expansion of deserts, for example. There could be severe heat waves, decreased crop yields and extinction of species.

It is important for us to discover ways to implement new technologies that can help capture carbon dioxide and even convert it into something more useful to reduce global warming effects. Once such technology is a nanostructured membrane that is being researched for use in coal-fired power plants. A coal-fired power plant can produce approximately 500 tons of carbon dioxide every hour. Research scientists are trying to create ways of being able to capture and store the carbon dioxide that comes from power plants using nanotechnology.

General Electric, for example, is investing in what is known as cleaner coal technology (also known as **integrated gasification combined cycle technology**, or IGCC) which involves turning coal into a gas and removing the carbon dioxide before it gets burned. One technology being researched for removing undesirable components in coal is a membrane with nanoscale pores. A **membrane** is a physical barrier that will allow certain gases to penetrate its surface much more quickly than others, depending on pore size. The pores are on the same scale as individual gas molecules in this case. With the correct nanoscale engineering of the membrane's structure, scientists are able to selectively transport different gases along the membrane walls. It is possible to produce a number of different membranes with different pore sizes to favor larger or smaller gas molecules. Current research involves making membrane units that are large enough to be able to filter out carbon dioxide at the large scales produced by coal-using power plants.

In addition to capturing carbon dioxide from power plants before it combusts, there is ongoing research focused on turning carbon dioxide into something useful. At the University of Georgia in Athens (UGA), researchers have found a way to transform the carbon dioxide trapped in the atmosphere into useful biofuels. Recall from earlier in the chapter that plants use sunlight to transform water and carbon dioxide into sugars that they need for life-sustaining energy. This process is known as **photosynthesis**.

The research group at UGA has essentially created a microorganism that mimics plant photosynthesis. The sugars made by plants can be fermented into fuels, such as ethanol, but the process is fairly inefficient, as it can be very difficult to remove the sugars from within the plants' cell walls. This new process will essentially eliminate the need to remove sugars from plants, and be able to generate the useful fuel directly.

The process uses a microorganism called *Pyrococcus furiosus*. This particular organism lives in the super-heated ocean waters that exist near geothermal vents where carbon dioxide is released from within the earth's crust. *Pyrococcus furiosus* uses the carbon dioxide as plants do, turning it into sugar. UGA has altered the genetic make-up of the organism so that it can feed on carbon dioxide at much lower temperatures. The organism turns carbon dioxide into 3-hydroxypropionic acid, which is a commonly used industrial chemical used in polymer synthesis and other manufacture of products.

As with the membrane technology, research is ongoing, and being scaled up to handle larger quantities of carbon dioxide.

Gasoline is another source of carbon dioxide in our atmosphere along with the burning of coal. Gasoline is one of the most commonly used fuels for powering automobiles. It is a hydrocarbon derived from petroleum, which is a non-renewable resource. When gasoline is burned, it produces a large amount of heat as the hydrocarbons are broken down into smaller molecules. One of the main by-products of the combustion of gasoline is the greenhouse gas carbon dioxide.

There is a great deal of interest in using catalysts to turn the carbon dioxide emitted from cars back into larger hydrocarbons or another more useful fuel. A **catalyst** is a material that increases the rate of a chemical reaction by lowering the amount of energy required for the reaction to progress. Unlike the starting materials in a chemical reaction, the catalyst is not consumed. In other words, after the catalyst has been used, it can be recovered and used again in subsequent chemical reactions. Nanoscale materials make more effective catalysts as they have a large surface area-to-volume ratio, and chemical reactions occur at a materials surface.

Copper is an example of a material being used as a catalyst in recycling carbon dioxide. It has the ability to reduce the carbon dioxide gas generated from automobiles or power plants into methane or methanol. Instead of carbon dioxide simply being emitted into the atmosphere, carbon dioxide could be circulated through a copper catalyst and turned into methane. Methane can then be used as a fuel.

Since copper can be easily oxidized, copper nanoparticles have been coated with gold. These nanoparticle composites are much more stable than copper alone and are highly corrosion and oxidation-resistant. Research has also been conducted using iron, and nickel-based catalysts to successfully convert carbon dioxide into two, three, and four-carbon long molecules.

While it is a main contributor to global warming, carbon dioxide is not the only greenhouse gas that exists. In addition to carbon dioxide, gases such as nitrous oxide (NO_2), hydrofluorocarbons (HFCs), perfluorocarbons (PCSs), sulfur hexafluoride (SF_6), and ozone (O_3) also trap heat effectively. Nitrous oxide is a far more powerful greenhouse gas than carbon dioxide. It is naturally emitted from oceans and from soil, but nitrogen fertilizers and fossil fuel combustion also contribute to NO_2 emission into the atmosphere.

Hydrofluorocarbons only make up a small portion of the total atmospheric greenhouse gas content, but are extremely potent. Certain types of HFCs can be 20,000 times more efficient at absorbing infrared radiation. HFCs are often used in refrigeration and air conditioning, as well as in aerosols and various industrial solvents. Perfluorocarbons are used in semiconductor manufacturing, and last in the atmosphere for up to 50,000 years.

Sulfur hexafluoride is the most potent of greenhouse gases. The Intergovernmental Panel of Climate Change identified it has being almost 24,000 times more potent than carbon dioxide. It is also used in semiconductor manufacturing, as well as in certain sneaker and tire manufacturing, for example.

Ozone occurs both naturally and can also be cause by human activities. While it is a necessary component in the upper atmosphere sine it protects the earth from harmful wavelengths of ultraviolet radiation, in the lower atmosphere it is the main component of smog.

As clean-air laws become stricter to control these and other harmful chemicals in our atmosphere, more research is using nanoscale materials to help clean our air. One such example is an air filtration system using porous manganese oxide decorated with gold nanoparticles as catalysts. This unique system has been shown to very effectively remove harmful volatile organic compounds as well as the nitrogen and sulfur-based pollutants. The large surface area of the small scale pores and gold nanoparticles provides a great number of sites onto which the gases can adsorb. Once adsorbed onto the surface, the catalyst allows of a very efficient break-down of the chemicals.

Other air purification systems are now commercially available. The NanoBreeze, for example, uses titanium dioxide nanoparticles and UV light to clean air. The titanium dioxide nanoparticles are able to easily oxidize gaseous pollutants when activated with ultraviolet wavelengths of light. They are able to destroy biological pollutants as well, such as pathogenic viruses and bacteria.

10.2 CLEANER WATER

Many of our natural bodies of water have become contaminated from human influences. Pollution of lakes, rivers, oceans and groundwater is a very serious global problem. Several sources have suggested that **water pollution** is the leading worldwide cause of death and disease, and it is estimated that 14,000 people die every day as a direct result of polluted water. Water can be defined as polluted if it contains contaminants that do not support human use or impair its ability to support biotic communities such as fish and aquatic plant life.

There are many factors that can contribute to water pollution. One such contributor is the class of organisms known as **pathogens**, or disease-causing organisms. High levels of pathogens are often the result of sewage discharges that have not been properly treated. Many bacterial infections such as E. coli, cholera, botulism, dysentery and Legionnaires' disease result from drinking of contaminated water. Protozoa, or one-celled organisms also account for many intestinal infections from water, such as amoebas. There are also viral infections like Hepatitis A and polio that can result from the consumption of polluted water.

In addition to pathogens, chemical species can cause water to become polluted. Insecticides, herbicides, detergents, hydrocarbons, oil, petroleum, industrial solvents, and fertilizers, and acids are just a few types of compounds that contribute to global water pollution.

So how can nanotechnology help to clean up contaminated water?

There are several nanotechnology-based solutions for cleaning up oil spills. The recent oil spill in the Gulf of Mexico reminded us of the catastrophic environmental impact an oil spill can have on an environment. It is not the first example of a large-scale oil spill, and sadly will likely not be the last. **Aerogels** are one technology that may be used for oil spill response. As was mentioned in Chapter 6, aerogels are very porous and ultra-lightweight materials that are made by removing the liquid component from a gel. Researchers have made a clay-based aerogel that is capable of absorbing oil within its vast pore network. Since the absorption is physical, meaning no chemical change takes place when it is absorbed in the aerogel, it can in theory be recycled once cleaned from a spill site.

A similar idea involves using carbon nanotubes in a 'sponge' form for absorbing oil. A research project was devised wherein boron atoms were selectively substituted into the pure carbon nanotube lattice. When there is a different atom in the lattice, its perfect network is disrupted and the tubes grow into curved structures that result in a woven sponge-like material. Carbon nanotube sponges have been shown to absorb up to 100 times their weight in oil. Again, since no chemical change takes place, the oil can be salvaged and the sponge can be used over and over again. If iron is added to the carbon nanotube sponge it results in having some magnetic character. Magnetic carbon nanotube sponges can easily be controlled and removed with strong magnets in a real-life oil cleanup scenario.

Nanotechnology may also be used to remove pathogens from drinking water. Currently, water treatment plants clean pathogens from water using a number of different techniques. These include chemical treatment with chlorine or iodine, UV treatment, and various filtration membranes. While large-scale water treatment plants are a common part of the infrastructure in industrialized countries, there are many situations where centralized water treatment is not an option. In this case, **point-of-use** water treatment is required, for example in poor or remote areas that don't have centralized water treatment facilities.

Point-of-use water treatment refers to purifying water where it will be consumed, and should be simple, effective, inexpensive, and require little or no power. Nanotechnology provides many solutions for potential point-of-use water treatments.

Carbon nanotubes are a great candidate for this application as well. Many organic materials such as pathogens have a high affinity for carbon nanotubes. Such pollutants could be removed from contaminated water via carbon nanotube filters.

As was additionally mentioned in Chapter 6, silver has very unique antimicrobial properties. A new water filter has been created that employs cotton decorated with silver nanowires. It works very quickly and not only does it block bacteria from getting through the filter, but also kills the pathogens in the process.

In places that do not have centralized water treatment facilities, or do not collect sufficient fresh water, but do have access to salt water, nanotechnology may also offer solutions for desalinating water. **Desalination** refers to any process that aids in removing salt from water. While 70% of the Earth's surface is covered in water, most of it is not suitable for human consumption. Salt water is not potable because our kidneys can only make urine that contains a lower concentration of salt than is contained in natural sea or ocean water. In other words, to remove all of the excess salt from your system, you would have to urinate more than you drank, therefore causing severe dehydration.

Researchers have now shown a new method for desalination of water using the **ion concentration polarization** phenomenon (ICP). ICP is a phenomenon that occurs when an ion current is passed through an ion-selective membrane. A team at MIT has created a small chip based on ICP that can remove salt from water. Their device contains two electrodes, one positive and one negatively charged. The negative ions get attracted to the positive electrode and the positive ions get attracted to the negative electrode. As water goes through the device, the ions from salt (for example, sodium ions and chloride ions) attach to the electrodes, leaving only pure water. Using nanoscale materials allows more surface area to which ions may be attracted

Carbon atoms are a main component in all living organisms, and are also part of the ocean, air, rocks, etc. In the atmosphere, carbon atoms can exist in molecules of carbon dioxide, or CO_2. Plants use these CO_2 molecules along with water and sunlight to form life-sustaining carbohydrates. When plants die, they decay and the carbon may turn into other sources of fuel such as coal or oil. When humans burn the coal or oil, CO_2 is released back into the atmosphere.

NANOSPEAK - WITH ANDREW WEST (CHAPTER 10)

1. Where do you work and what is your job title?

   I work at GlobalFoundries as a Senior Technician. My job involves using a tool called a CLM - it is a sample prep tool that houses a scanning electron microscope and a focused ion beam gun.

2. What is your educational background?

   I am working on an AAS degree in nanotechnology.

3. How did you get interested in nanotechnology/where did you hear about it?

   I had an old coworker who went through a two year community college nanotechnology program and got a great job at GE. With the nanotechology industry growing around the world I felt there was a lot of potential in the field. I also like technology – it has always interested me.

4. What advice would you give to someone who might want to get into the field?

   The nanotechnology industry is quite large with a number of different sub-industries included. You can work for semiconductor manufacturers, materials engineering, and energy companies, to name a few. I think it's wise to learn a little about each so you have more of an opportunity for employment. I feel like my community college program helped me see the different industries from the start.

CHAPTER 10 SUMMARY

- Carbon dioxide exists naturally in our atmosphere but humans contribute a large amount as well.
- Carbon dioxide absorbs infrared frequencies and radiates them back to earth causing a warming effect. Too much carbon dioxide in the atmosphere can cause damaging global climate change.
- Nanomaterials can be used in various ways to sequester excess carbon dioxide from the atmosphere or prevent its entry into the air.
- Water pollution is a serious global issue causing thousands of deaths daily.
- There are steps being made to use nanotechnology to clean polluted water for example by killing infection bacteria or removing oil.

REFERENCES

1. D. Archer, *The Global Carbon Cycle*, Princeton University Press, 2010.
2. D. Williams, "GE Introduces Next Generation Clean Coal Technology," *Power Engineering International*, October 2013.
3. Y. Han, A. Hawkins, M. Adams, "Impact of Substrate Glycoside Linkage and Elemental Sulfur on Bioenergetics of and Hydrogen Production by the Hyperthermophilic Archaeon Pyrococcus furiosus," *Applied Environmental Microbiology*, **73 (21)**, 2007, p. 6194.
4. J. Graciani, K. Mudiyanselage, F. Xu, A. Barber, J. Evans, S. Senanayake, D. Stacchiola, P. Liu, J. Hrbek, J. Sans, J. Rodriguez, "Highly Active Copper-Ceria and

Copper-Ceria-Titania Catalysts for Methanol Synthesis from CO_2," *Science,* **345,** 2014, p. 546.

5. N. Cioffi, M. Rai, *Nano-Antmicrobials: Progress and Prospects,* Springer, 2012.

6. Q. Zheng, Z. Cai, S. Gong, "Green Synthesis of Polyvinyl Alcohol-Cellulose Nanofibril Hybrid Aerogels and Their Use as Superabsorbents," *Journal of Materials Chemistry A,* **2,** 2014, p. 3110.

7. X. Gui, J. Wei, K. Wang, A. Cao, H. Zhu, Y. Jia, Q. Shu, D. Wu, "Carbon Nanotube Sponges," *Advanced Materials,* **22 (5),** 2010, p. 617.

8. S. Kim, S. Ko, K. Kang, J. Han, "Direct Seawater Desalination by Ion Concentration Polarization," *Nature Nanotechnology,* **5,** 2010, p. 297.

CHAPTER 11   MICROELECTRONICS

KEY TERMS

DISCRETE DEVICE

TRANSISTOR

RESISTOR

CAPACITOR

DIODE

INTEGRATED CIRCUIT DEVICE

CHIPS

INSULATOR

SEMICONDUCTOR

BAND GAP ENERGY

P-TYPE DOPANTS

N-TYPE DOPANTS

DIELECTRIC MATERIAL

LIGHT-EMITTING DIODE

SILICON-CONTROLLED RECTIFIER

BIPOLAR JUNCTION TRANSISTOR

METAL OXIDE SEMICONDUCTOR FIELD
    EFFECT TRANSISTOR

COMPUTER PROCESSOR

BINARY SYSTEM

BIT

BYTE

MOORE'S LAW

EXTREME ULTRAVIOLET LITHOGRAPHY

X-RAY LITHOGRAPHY

ELECTRON BEAM LITHOGRAPHY

NANOELECTRONICS

MOLETRONICS

GROW-AND-PLACE

GROW-IN-PLACE

INTRODUCTION

In Chapter 11, you will be introduced to the importance of nanotechnology in the field of microelectronics. There is a rich history regarding semiconductor-based electronics and the shrinking of basic electronic devices. You will learn about discrete devices such as resistors, capacitors, diodes, and transistors and how they come together to make integrated circuit chips. New trends and future possibilities for the continued reduction of size in electronics will also be outlined in this chapter.

Some things to think about:

- Why do we use silicon in electronic devices?
- What makes a material semiconducting as opposed to insulating or conducting?
- Why is nanotechnology so important in microelectronics?
- What are some new trends in lithography and why do we need to move away from our current lithographic methods?

11.1   HISTORY

The era of semiconductor electronics began in the 1940s. Two scientists, American physicist and electrical engineer John Bardeen, and American physicist Walter Brattain made a solid state electrical device out of germanium. They noticed that when electrical signals were applied to contacts on a piece of germanium, the amount of output power was greater than in amount of input power. This was the first example of a **transistor**, and was made at Bell Laboratory. Their supervisor, American physicist William Shockley, also worked to find out how a bipolar transistor functioned. The three scientists shared the 1956 Nobel Prize in Physics for the invention of this transistor device.

The demand for electronic devices only grew in as time progressed into the 1950s. Still using germanium, transistor-based devices began to replace vacuum tubes in equipment because they were smaller, faster, and more efficient. Germanium was then replaced by silicon as the semiconductor material of choice. The industry at that time produced many **discrete devices** such as resistors, capacitors, diodes, and transistors. These only contained one device per piece.

Towards the end of the 1950s, American engineer Jack Kilby at Texas Instruments realized that discrete devices could be made on the same piece of silicon. The discrete devices could then be connected together to form a circuit. He used a piece of germanium with one transistor built onto it already, and added a capacitor and three resistors. Once the transistor, capacitor, and three resistors were connected with platinum, the first **integrated circuit (IC) device** was invented. An IC device contains multiple discrete devices connected on the same piece of semiconducting material, and are now commonly referred to as **chips**.

Around the same time that Jack Kilby was working at Texas Instruments, America physicist Robert Noyce at Fairchild was working on a similar idea. Instead of using platinum wires to connect the components, Robert Noyce's IC device used lines of etched aluminum on the surface for device interconnections. Additionally, he used silicon instead of germanium. Silicon is the preferred semiconductor material because it can easily be oxidized to form an insulating silicon dioxide material directly on its surface. Fairchild made the first commercially available IC devices in 1961. They were around 10 mm in size, contained four transistors and cost $150.00. These devices served as a model for all current IC devices.

11.2 BASIC DEVICES

Recall that semiconductors are materials with electrical conductivity somewhere between a good insulator and a good conductor. The most commonly used semiconducting material is silicon, which makes four covalent bonds to other silicon atoms.

At absolute zero, all of the covalently bonded electrons cause the material to act as a perfect insulator. As temperature increases, however, electrons may become mobile. The fundamental difference between a material which behaves as an insulator and a semiconductor lies within its band gap energy, or the amount of energy required to excite an electron from the valence band into the conduction band. Most metals have a band structure where the valence and conduction band overlap or are separated by only a small amount of energy so that they can be excited at room temperature energy. Recall that for insulators, the band gap is so large that electrons cannot be excited from the valence band into the conduction band. Semiconductor band gaps lie somewhere in the middle so that there are some small number of electrons that can be excited into the conduction band at room temperature.

Remember from Chapter 3 that we can control the electrical character of a semiconductor such as silicon by intentionally adding dopants. **P-type dopants** contain three valence electrons (for example boron) and create holes within a semiconductor band gap. **N-type dopants** contain five valence electrons (like phosphorus or arsenic) and add an extra electron into the valence band. The higher the concentration of dopant atoms, the lower the resistivity of silicon will be.

The basic devices used on silicon in an IC chip are resistors, capacitors, diodes, transistors, and metal oxide semiconductor field effect transistors (MOSFETs).

11.2.1 RESISTORS

The electric current we use is essentially just electrons flowing from one place to another. A **resistor** is a device that resists how much electrical current, or how many electrons, can flow through a device. Resistors are an important part of integrated circuits, as they can help a circuit achieve different functions. For example, resistors can help reduce voltage to part of a circuit, control how much voltage goes from one component to another or simply limit the amount of current another component of the circuit receives.

The formula for resistance of a resistor is shown in **Equation 11.1.**

Equation 11.1 $R = \rho \dfrac{l}{wh}$

In this equation, R is the resistance to movement of electrons, ρ is the resistivity, *w* is the width, *l* is the length, and *h* is the height.

Previously in IC manufacturing, doped silicon was used to make resistors. Now, the most commonly used material is polycrystalline silicon, or polysilicon. Polysilicon has ordered areas known as grains, but overall is not a single perfectly ordered crystal.

11.2.2 CAPACITORS

Capacitors are also very important components in IC chips. Essentially, a **capacitor** is composed of two conducting plates separated by a **dielectric material**. A dielectric material is an electrical insulator. The equation for the capacitance of a capacitor, or how much charge it can hold is **Equation 11.2.**

Equation 11.2 $C = \kappa \varepsilon_0 \dfrac{hl}{d}$

where C is capacitance of the capacitor, ε_0 is the electric constant (8.85 x 10^{-12} F/m),

κ is the dielectric constant of the insulating layer, *h* and *l* are the height and length of the conducting plate, and *d* is the distance between the two conducting plates.

Capacitors have many functions, for example in random-access memory chips to store electrical charges and maintain computer memory. Capacitors can also be used to keep voltage at a constant level, block direct current (DC) and adjust frequencies. An IC chip generally employs polysilicon as the conducting plate part of the capacitor. The dielectric material can vary, but some materials that are often used are silicon dioxide, silicon nitride, and titanium dioxide. As IC chip feature size shrinks, dielectric materials with higher capacitances are needed to hold the same amount of charge on a much smaller size scale. There are many different styles of capacitors that can be used in an IC chip, for example, planar, stack, or deep trench styles.

11.2.3   DIODES

The prefix "di-" means two. A **diode** has two terminals, each with a high resistance to current flowing in one direction and a low resistance to current flowing in the opposite direction. In other words, a diode only allows electric current to go through in one direction when voltage is applied. Diodes were briefly discussed in Chapter 9. When a p-type semiconductor is placed adjacent to an n-type semiconductor, some of the electrons from the n-side flow into the p-side, and some of the holes or positive charges from the p-side flow into the n-side. The area where this exchange occurs is called the transition region, or P-N junction. What results is a P-N junction diode when voltage is applied.

The voltage at this transition region can be written according to the following equation:

Equation 11.3 $V_o \frac{kT}{q} = l_n \frac{N_a N_d}{n_i}$

In this equation, *k* is the Boltzmann constant, *T* is the temperature, *q* is the charge, N_a is the concentration of p-type dopant (acceptor dopant), N_d is the concentration of n-type dopant (donor dopant), and n_i is the intrinsic carrier concentration.

There are a number of applications that use diodes, and diodes can fall into different categories. In general, diodes are used to prevent damage to other polarized components within a circuit. In other words, a diode can protect against current flowing in the wrong direction. Different types of diodes are used frequently in common devices. **Light-emitting diodes** (LED) emit visible light in a variety of colors. **Silicon-controlled rectifiers** (SRC) are switches that control current that are usually used in light dimmer switches. A rectifier transforms alternating current (AC) to provide DC current.

11.2.4 TRANSISTORS

A transistor is a device that controls the movement of electrons in a circuit. Essentially, they work like a water faucet that can stop electrons from moving all together, but can also control the number of electrons moving, or the amount of current.

If a p-doped layer of silicon is adjacent to an n-doped layer of silicon, we know that a diode will result. To create a transistor, several doped layers must be placed back to back in a configuration of p-type, n-type, p-type (PNP) or n-type, p-type, n-type (NPN). The point of contact of the three is called a **bipolar junction transistor.**

A bipolar transistor is generally used for its ability to amplify electrical signals. These transistors dominated the semiconductor industry from the 1950s to the 1980s. Today, metal-oxide-semiconductor field-effect transistors are the prevalent type of transistor used in IC devices.

MOSFET stands for **metal-oxide semiconductor field-effect transistor**. They contain a source, or a place where electrons enter, a drain from which electrons exit, and a gate which can open and close. When the gate is closed, electrons can flow from the source to the drain. When the gate is in the open position, the little switch is 'off,' where electrons cannot flow out of the source. The on and off positions create the 1's and 0's used in electronic devices for data.

11.3 WHY NANOTECHNOLOGY IS CRITICAL TO MICROELECTRONICS

A computer's processor is an integrated circuit chip. The speed of a computer's processing will rely on the size of the processor chip. In general, the smaller the device circuitry, the faster it can process information.

Computers perform functions based on a system of 1s and 0s known as the **binary system**. The 1s and 0s correspond to a transistor 'switch' being turned on or off. 1s and 0s are combined in a multitude of patterns in order for a computer to solve problems.

A **bit** is a single 1 or 0, and stands for **b**inary dig**it**. A **byte** is composed of 8 bits. This term is used when describing the computer's ability to store information. A megabyte (MB) is approximately one million bytes. Technically, since the system is binary, a megabyte is 2^{20} bytes, or 1,048,576, but this number is generally rounded for convenience. A gigabyte (GB) represents a billion bytes, and a terabyte (TB) is approximately one trillion bytes.

Again, a transistor acts as a switch that dictates whether the circuit is on or off, or a 1 or 0. Transistors get combined in chips to generate processors. It takes many millions of transistors to make our current processors as small and fast as they are. A smaller transistor means a smaller switch, which in turn results in faster switching between on and off position.

Moore's Law, named for Gordon Moore, co-founder of Intel, is an observation that over the history of computer processors, the number of transistors on an IC chip doubles about every 18 months to 2 years. His prediction was fairly accurate from 1958 until recently. It is estimated that after 2013, transistor counts will double only every three years. In the 1970s, processors contain between 4,000 and 8,000 transistors. The latest processors contain more than 2 billion!

It is very technologically challenging to create such small feature sizes, and silicon may be reaching its physical limits. The lithography process, outlined in Chapter 7, is the most fundamental step used to pattern silicon wafers. Ultraviolet photolithography will run its course, as the wavelength of ultraviolet light is larger than the desired feature sizes of 7 nm, and soon to be even smaller. Alternative lithographic techniques will be required, for example, extreme ultraviolet (EUV), X-ray, or electron beam lithography.

Extreme ultraviolet, or EUV is electromagnetic radiation with wavelengths between 11 and 14 nm. The range of the electromagnetic spectrum between 1 and 50 nm is the overlapping region between ultraviolet and x-ray, named both extreme UV and soft x-ray. The lithographic technique was named EUV so as to avoid confusion with x-ray lithography. By using such short wavelengths, and reducing the aperture through which the light passes, lithographic resolution can be increased. Since materials will strongly absorb such short and energetic wavelengths, no materials known can be used as EUV lenses. For this reason, EUV systems are mirror-based.

X-ray lithography employs wavelengths that are less than 5 nm in length. Since the wavelengths of x-rays are shorter than ultraviolet rays, they can form smaller feature sizes with higher resolution than traditional ultraviolet methods. There aren't materials that reflect x-rays, so mirrors cannot be used in this process as they can in EUV. Instead, X-ray lithography is a direct printing process, where the x-rays shine through a mask and expose the photoresist on the surface. In order to use this method in a high volume manufacturing setting, all of the lithography exposure systems would need to be redesigned. X-rays cannot be focused, so the mask and wafer feature size need to be exactly the same. Additionally, the mask needs a thicker gold layer to block unwanted x-rays from reaching the photoresist surface.

Electron beam lithography, or e-beam lithography, employs electrons, which can behave as both particles and waves. The wavelength of an electron beam can be calculated according to **Equation 11.4.**

Equation 11.4 $\lambda = \frac{1.23}{\sqrt{V}} \text{nm}$

Therefore, an electron beam operating at 50 kV will have a wavelength of approximately 0.006 nm. This is much shorter than the wavelength of ultraviolet light. Like ultraviolet light, an electron beam can be reflected and focused by electric and magnetic fields.

Early e-beam lithography systems used a scanning beam of electrons to draw patterns directly into photoresist. The solubility of an electron-sensitive photoresist will change when it comes into contact with an e-beam. The process of writing directly into photoresist using an e-beam can be somewhat slow for high volume manufacturing.

11.4 FUTURE TRENDS IN MICROELECTRONICS

Feature size is expected to continue shrinking past the physical size limit that top-down manufacturing permits. Additionally, the costs associated with new lithographic techniques and the many processing steps involved in IC manufacturing are becoming prohibitively high. Some alternative manufacturing methods will need to be explored to continue producing our electronic devices.

It has been suggested that top-down nanofabrication should be abandoned and there should be a move toward bottom-up or some hybrid fabrication techniques. Some approaches that have been proposed are microelectronic devices based on individual nanoparticles, or **nanoelectronics**, and microelectronics that are based on individual molecules termed **moletronics**.

The advantages of these new techniques would lie in the inherently small size of molecules and nanoparticles. Since nanoparticles and molecules are already small and can potentially self-assemble, the need for lithography and etching is eliminated. The physical and chemical properties of materials on such a small scale are very unique, so there are the possibilities for new types of devices with a variation of device physics and device chemistry.

The building blocks for nanoelectronics are individual nanoparticles such as nanowires, nanotubes, or quantum dots. In Chapter 7, we discussed some of the ways in which individual nanoparticles for bottom-up fabrication can be synthesized. Making devices out of nanoparticles can be done in one of two ways, either by a grow-and-place method or a grow-in-place method.

A **grow-and-place** approach involves first growing the nanoparticles, for example nanowires or nanotubes. Once the nanoparticles are grown, they can be harvested, placed on a substrate, and a device can then be made. Carbon nanotube field effect transistors, for example, have been made via the grow-and-place method. A carbon nanotube (CNT) field effect transistor uses a single carbon nanotube to channel electrons as opposed to bulk silicon. The CNTs are grown, suspended in a solution, and manipulated into a small discrete device.

The **grow-in-place** approach to nanoelectroncs involves synthesizing the individual nanoparticles into the device where it will be used. This approach has been used successfully to grow silicon nanowires in place to generate a silicon nanowire field effect transistor.

Moletronics is another potential alternative to the microelectronics industry. Moletronics is the concept that individual molecules can be used at the components in IC devices. Individual molecules make attractive candidates from electronic devices due to their small size. Currently, top-down manufacturing produces features that are on the order of 40 nm. Molecules are about 100 times smaller than these smallest of printed features.

Specially synthesized self-assembling molecules can exhibit semiconductive properties that give them the ability to hold a charge or behave like switches or memory. This means molecular electronics have the potential to replace the transistors, diodes, and conductors of our current silicon-based IC devices.

Research has shown that molecular devices made with transistors can be produced cheaply in huge numbers, will be able to compute faster, remember for longer amounts of time, and consume a very small amount of power. Additionally, moletronics have the potential to go beyond the limitations of our current information storage technologies, providing memory systems that are so powerful, small, and inexpensive that the entire Internet could potentially be reserved on a single desktop.

While moletronics research is still in its infancy, it is clear there is an immense amount of potential for the future of electronics made from individual self-assembling molecules.

NANOSPEAK – WITH DEREK GRIPPIN (CHAPTER 11)

1. Where do you work and what is your job title?
   My title is "Research Technician II" at the College of Nanoscale Science and Engineering.

2. What is your educational background?
   I have an applied associate's degree in Nanoscale Materials Technology.

3. How did you get interested in nanotechnology/where did you hear about it?
   After serving as an Avionics technician in the marines, the V.A. office directed me to a Nanoscale Materials Technology program. The more classes I took, the more I liked it!

4. What advice would you give to someone who might want to get into the field?
   I would say learn as much as you can from anyone who is willing to teach you! Keep an open mind about continuing your education as well.

CHAPTER 11 SUMMARY

- The era of semiconductor electronics began in the 1940s when John Bardeen and Walter Brittain made a solid state transistor out of germanium at Bell Labs.
- Transistor-based devices began to replace vacuum tubes, and germanium was replaced by silicon.
- In the 1950's, Jack Kilby made the first integrated circuit device, with several discrete devices connected on the same piece of semiconductor material. At the same time, Robert Noyce made a similar device with silicon and aluminum connecting the discrete devices.

- Silicon is an important material since it can conduct at room temperature, and can be doped to tailor its electrical behavior.
- The basic devices used in integrated circuits are resistors, capacitors, diodes, and transistors.
- A resistor is a device that resists current flowing through a material. They are generally made out of polysilicon.
- Capacitors hold charge and are essentially made of two conducting plates separated by an insulating material called a dielectric. Usually polysilicon acts as the conductor, and the dielectric is a material such as silicon dioxide or titanium dioxide.
- Diodes are devices that only allow current to flow through in one direction. They are created by placing p-type silicon adjacent to n-type.
- Transistors are devices that control the movement of electrons in a circuit. Electronic devices often contain metal-oxide semiconductor field-effect transistors (MOSFETs) that act as little switches which can be turned on and off to process information.
- Computer processors are getting smaller and faster, with more transistors per device. As circuitry needs to get smaller, new trends are being developed to print such nanoscale features, for example extreme ultraviolet, x-ray, and electron beam lithography.
- Research is being conducted to find ways of employing bottom-up techniques for creating small features.

REFERENCES

1. M. Quirk, J. Serda, *Semiconductor Manufacturing Technology*, 1ˢᵗ *Ed.*, Prentice Hall, 2000.
2. H. Xiao, *Introduction to Semiconductor Manufacturing Technology*, Prentice Hall, 2000.
3. K. Kwok, J. Ellenbogen, "Moletronics: Future Electronics," *Materials Today*, **5 (12)**, 2002, p. 28.

CHAPTER 12   ADDITIONAL APPLICATIONS OF NANOTECHNOLOGY

KEYTERMS

CATHODE RAY TUBE	NANOFLUID
PIXEL	NANOCOMPOSITE
LIQUID CRYSTAL DISPLAY	LIPOSOME
QUANTUM DOT	SOLID LIPID NANOPARTICLE
ORGANIC LIGHT EMITTING DIODE	NANOSPHERE
ELECTRONIC PAPER	NANOEMULSION
ELECTRONIC INK	MICROELECTROMECHANICAL SYSTEMS
LOTUS EFFECT	ACTUATOR
SUPERHYDROPHOBIC	SENSOR
NANOFABRIC	THERMOPILE

INTRODUCTION

As has been outlined in the previous chapters, there are many exciting applications of nanotechnology in such fields as biology, medicine, energy, and electronics. Additionally, there is the potential for use of nanoscale materials in every day consumer products. In this chapter, a small cross-section of the nearly infinite uses of nanoscale materials will be discussed. Some areas that may greatly benefit from the use of nanoscale materials are display technology, textiles, sporting equipment, automobiles, cosmetics, home goods, and the food and beverage industry.

Some things to think about:

- What are some other industries that may employ nanotechnology for product improvement?
- Do you own any products that contain nanomaterials?
- What are your own ideas for implementation of nanotechnology in everyday life?

1.1   DISPLAY TECHNOLOGY

Older display technology contained cathode ray tubes. A cathode ray tube is a vacuum tube with an electron beam source and a fluorescent screen that can be used to view images. The electron beam hits individual pixels, or small elements of an image, one after another causing them to glow. This process happens so quickly that the image appears continuous to our eyes, and the flickering is not noticed.

Cathode-ray tube displays are very large, bulky, hot, and inefficient. They have largely been replaced by newer, small, and more efficient technologies, for example, liquid crystal displays (LCDs). Liquid crystal screens are energy efficient and can be easily powered with batteries. They are also easier to dispose of once no longer working.

While liquid crystal display technology is popular, it is not without room for improvement. The images are not as bright as they could be, and depending on viewing angle, the image may not be seen properly. Nanotechnology is helping to make displays that are even thinner, brighter, and have better contrast between hues.

An improvement to liquid crystal display technology is the use of quantum dots. As discussed in Chapter 6, quantum dots can be defined as nanoparticles made of semiconducting materials. The unique optical properties of quantum dots allows for a display that yields approximately 50% more color than a traditional liquid crystal display. The quantum dot displays allow light to be supplied on demand, making them not only vibrant in color but highly efficient, with devices that have markedly longer battery lives.

An organic light-emitting diode (OLED) is a light-emitting diode in which the electroluminescent layer is an organic, or carbon-based film that emits light when an electric current is applied. There are two families of OLEDs. One type of OLED is based on small organic molecules, while the other type employs long-chained polymer molecules. What is unique about OLED displays is that they do not need to be backlit. This means the displays can show deep black hues and can be incredibly thin and light. Polymer LEDs allow for displays that are thin and flexible as paper. They use a thin layer of plastic polymer a few nanometers thick between two thin layers of metal. When an electric current passes through the plastic, it glows and the result is an image on the plastic screen. In addition to being lightweight and flexible, OLEDs will be inexpensive to manufacture, allow for a wide range of viewing angles, have improved brightness, and will have faster refresh rate than liquid crystal displays.

Another alternative to liquid crystal display technology includes electronic paper with electronic ink. Electronic paper is a paper thin flexible screen that you can write on and use to view documents. While it is still in prototype stage, one idea to make electronic paper is by using electronic ink, which has been used to create e-book readers. Electronic ink's exact composition is proprietary, but it consists of tiny black and white spheres that float up and down in liquid when electrically charged to form an image. While Electronic ink is still black and white, and microscale, it is a first step toward real electronic paper.

1.2   FABRICS AND CLOTHING

The textile industry has benefitted greatly as a direct result of advancements in nanotechnology. Nanotechnology can be used to enhance many properties of fabrics and yield textiles that are softer, more durable, breathable, fire retardant, and antimicrobial. More traditional fabrics are made of cotton. Cotton fibers can yield very desirable fabrics that are highly water absorbent, soft, comfortable, and breathable. The problem with cotton is that it isn't very durable after prolonged use. Additionally, it soils very easily, and is highly flammable. We also have synthetic fabrics that are made of polymer materials such as nylon and polyesters. While they tend to be more durable than cotton fiber fabrics, they lack the comfort and softness of natural cotton.

Nanoscale materials can be integrated into cotton or polymer fibers to obtain composite nanofabrics. Nanofabrics can be made by incorporating the nanomaterials into threads to form nano-fibers, or by applying solutions containing nanomaterials to traditional cotton or polymer fabric surfaces.

In Chapter 8, the lotus effect was discussed as an example of using inspiration from nature to create advanced materials, or biomimicry. The nanoscale surface features of the lotus leaf were used as a model for creating spill-resistant fabric. The lotus leaves are covered in waxy crystals that about 1nm in diameter. They form a superhydrophic coating on the leaf that causes water to form into beads with a very large contact angle. As the water beads run off, they take dirt and dust particles away with them. Scientists have used this same idea to create nanoscale polymer 'hairs' on fabric that prevents liquids from making contact with the fabric's surface. These fabrics are water-proof, as well as dirt- and stain-proof.

Silver's antimicrobial properties have been exploited for thousands of years. Silver nanoparticles can be easily synthesized and attached to cotton or polymer fibers to make antimicrobial fabrics. The silver nanoparticles do not wash away after laundering, and inhibit the growth of bacteria and fungus. Silver nanofabrics have been used in a number of applications, for example, odor-eliminating socks, hospital gowns and bed sheets, military and healthcare worker uniforms, and sporting apparel.

Carbon nanotube composite yarns have been synthesized as well. These yarns have extremely high strength and can tolerate high temperatures. Since carbon nanotubes can also conduct an electric current, there is the potential to use these yarns for many applications such as electronic textiles which are capable of protecting against electrostatic charges and absorbing microwave frequencies.

There are many new potential applications of nanofabrics in areas one might not necessarily find to be obvious. For example, nanofabrics have the potential to be used in tissue engineering as templates for the growth of organ tissue, bones, tendons and ligaments. Their structure can act as a sort of scaffolding to grow actual tissue, using either natural or synthetic polymers. There is also the potential to create porous nanofabrics and load them with drugs which can be applied to targeted areas as a new drug delivery system. The drug can pass from the nanofabric to the tissue via a controlled diffusion. The diffusion rate can be controlled by changing the structure and composition of the nanofabric.

1.3   SPORTING GOODS

Nanotechnology is advantageous for use in sporting equipment, particularly when one considers that nanomaterials can help make materials lighter, more flexible, and stronger.

Nanoscale materials have successfully been integrated into any number of sporting items, and many are already commercially available.

About ten years ago, nanotechnology made its commercial debut in the tennis world. There have been tennis balls created with what is known as 'double core' technology. The inner core of this tennis ball contains a coating made of 1 nm clay particles mixed into rubber. These clay nanoparticles create a sort of maze that slows down the movement of air across its surface. The result is a ball that stays inflated for a longer amount of time, making the ball playable for double the lifetime of a traditional tennis ball.

Nanoscale materials have also been used in tennis racquets. The frames of tennis racquets incorporate silicon dioxide nanoparticles into the composite material that contains carbon fibers. The silicon dioxide ceramic particles fill in the gaps between the carbon fibers and increase the strength of the racquet, giving the user more power when striking the ball. More recently, racquets have been made with basalt instead of carbon fibers. Basalt is rock formed when lava cools rapidly. It contains many small pores that result from gas bubbles contained in the lava, and is therefore lightweight yet still strong. Racquet makers claim fibers made of basalt reduce vibrations in the racquet when striking the ball which gives the player better control. Basalt is also significantly less expensive than carbon fiber.

Golf is another sport that has benefitted greatly from nanotechnology. Carbon nanotubes have been integrated into high end golf clubs, making the club head much stronger. With a stronger club head, energy can be more efficiently transferred to the ball, which in turn allows the ball to travel faster. Replacing titanium-based club heads with carbon nanotube composites makes them lighter and lowers the center of gravity. This helps the golfer achieve more power and accuracy.

In addition to tennis and golf applications, other sporting equipment is being manufactured with nanoscale materials as well. Bicycle parts are being made with carbon nanotubes to increase strength and decrease weight of the bikes. Novel ski wax made with nanoscale materials help to increase ski gliding across snow and increase maximum speed of skiers. Baseball bats made with carbon fibers only contain resin between the strands of carbon. By adding carbon nanotubes to the resin, it becomes much tougher and stronger. The result is a baseball bat that gives maximum hitting performance for batters.

12.4   AUTOMOBILE INDUSTRY

The Society of Automotive Engineers (SAE) identifies many promising automotive applications of nanotechnology. One main area in which nanotechnology can be helpful to the automobile industry is in the removal of waste heat. There is increasing focus on making smaller components for engines with less mass, but the traditional approach of simply increasing the area available for heat exchange with coolant such as ethylene glycol and water to manage increased heat loads is no longer working. The idea of using nano-fluids for cooling purposes has been received with good success. Nano-fluids contain nanoparticle suspensions of such materials as gold, copper, alumina, copper oxide, and silicon carbide. The increased thermal conductivity of nanoscale materials has made for significantly improved coolants.

Nano-composites are making their mark on the automotive industry as well. A nano-composite is a material that contains nanoscale materials mixed into a matrix of another material such as a polymer. Adding nanoparticles to standard materials helps to improve upon their properties, for example, mechanical strength or thermal conductivity. Polymers containing a small amount of nanoparticle material (2-6%) exhibit unique properties that are useful for automobiles. Many automobile makers have taken advantage of nano-composites, for example, in the automobile body, truck cargo beds, improved fuel tanks, and engine covers.

There are many other potential uses of nanotechnology in automotive applications. Dirt-repellent coatings for paints and windows can help improve durability. To help with visibility though windows and windshields, anti-fogging and anti-reflective coatings may be applied that use nanoscale materials. Wear resistant tires, scratch resistant paints, and high power interconnects for hybrid vehicles are just a few improvements that nanotechnology can make toward exciting, novel automobile applications.

12.5   COSMETICS

Currently, there are two main uses of nanotechnology in the cosmetics industry. The first is in sunscreens, which regularly contain either titanium dioxide or zinc oxide nanoparticles. Titanium dioxide and zinc oxide nanoparticles have exceptional ability to absorb ultraviolet radiation from the sun. It is estimated that over 300 sunscreens on the market today contain nanoparticles.

Delivery of active agents to cells is the second main use of nanotechnology in cosmetics. Liposomes and solid lipid nanoparticles, discussed in Chapter 8, are used as carriers of cosmetic agents to cells for applications as skin moisturizers and anti-wrinkle compounds. These lipid-based nanoparticles are particularly effective in delivery of active agents to skin because they can easily merge with the lipid bilayer of cell membranes, facilitating delivery of compounds to the cell which would otherwise not be absorbed. They are also non-toxic and biocompatible.

Encapsulating active ingredients in nano-spheres or nano-emulsions has become popular in the cosmetics industry to increase penetration into the skin. Companies have used polymer 'nanocapsules' to deliver vitamin A deep into the skin to prevent wrinkles, for example. A newly developed facial mask contains gold nanoparticles which are claimed to have anti-inflammatory and anti-oxidant properties. This mask is supposed to aid with tissue regeneration, restore elasticity to skin, and reduce signs of aging.

12.6 HOME APPLIANCES

Microelectromechanical systems, or MEMS, are small devices that exist on the microscale of 1-100 micrometers. They usually consist of a central microprocessor device and several other components that interact with the devices' surroundings. MEMS are made using standing top-down nanofabrication techniques. They can be separated into two general categories: actuators and sensors. Actuators can manipulate or move objects, while sensors detect changes, for example changes in speed.

MEMS are being integrated into many commonly used home appliances. There is a dishwasher that contains a MEMS soil-sensor. This sensor helps to determine the appropriate amount of water necessary to properly clean the dishes per load without being wasteful of water.

Refrigerators with automatic ice cube makers have been fitted with temperature sensitive MEMS called thermopiles. These sensors can detect when the water in an ice cube is completely frozen and can be dropped into the bin for use. Some cook tops also employ temperature sensitive MEMS to regulate temperature by monitoring the temperature of cookware being used. When the cookware is removed, the energy transfer stops and the burner automatically cools. Light bulbs can use a MEMS-based cooling device that decreases the temperature of the bulb, which saves energy, and allows the light bulb to last longer.

12.7 PET PRODUCTS

It is estimated that Americans spend over $50 billion on their pets in a given year. Appoximately 65% of American families have at least one pet in their home, and so it is not surprising that there is room for new innovations in pet products. Using nanoscale materials, companies are making pet products that are cleaner and healthier for both pets and their owners.

Dog beds and dog clothing are being manufactured by a company called Nano Pet Products, LLC. They use a certified technology known as NanoSphere ® which is impregnated into dog bed and clothing fabric. The technology inhibits odor-causing bacteria from spreading through the fabric. Typically, pet odors are the direct result of oil from the dog's coat getting on the fabric. Bacteria then attack this oil. Nano Pet Products' technology have a very high oil repellency. The nanoparticles in the fabric then reduce the growth and spread of odor causing bacteria.

12.8 FOOD AND BEVERAGE

Some of the world's biggest food producers want the food industry to benefit from nanotechnology. One example is keeping food fresh for longer periods of time. Meats and cheeses are often packaged in protective cellophane wrapping to prevent drying or spoiling.

Air permeating packaging is the main reason food dries or spoils quickly. Nanoscale platelet particles can be mixed into a composite material with plastic to form a sort of maze that takes air a longer time to traverse. This concept is similar to the clay particles described in tennis ball technology. The longer air takes to reach food products, the longer the food will stay fresh.

Similar technology has been implemented in glass beverage bottles. Clay nanoparticles shaped as discs slow air from entering into food packaging, but can also aid in preventing air from escaping. When carbonated beverages are bottled in glass bottles with embedded clay nanoparticles, the beverages stay carbonated for longer amounts of time, since the carbon dioxide that keeps beverages 'bubbly' takes longer to escape.

Another way in which food spoils is via exposure to bacterial such as *E. coli* and *Salmonella*. Using fluorescent nanoparticles like quantum dots, bacterial contaminants can be specifically targeted. Bacteria tagged with quantum dots can quickly and easily be detected using a hand-held fluorescent light. This makes visualizing harmful bacteria on food a facile procedure to protect consumers from potential foodborne illness.

NANOSPEAK – WITH ROBERT BAUER (CHAPTER 12)

1. Where do you work and what is your job title?
    GlobalFoundries. I'm a Failure Analysis Engineering Technician.

2. What is your educational background?
    I have an AAS degree in nanotechnology.

3. How did you get interested in nanotechnology/where did you hear about it?
    I have always been interested in science and more specifically electronics, but never had a real direction for learning more about them. With some assistance and guidance from family, I learned about nanotechnology and researched it. I was interested enough to do a tour my community college and learn more. At that point I knew that was the field I wanted to be in.

4. What advice would you give to someone who might want to get into the field?
    There are so many different things that can be done in the field. Do your research and learn in what direction you want to go. Then, you can make a game plan for getting yourself there. It will not always be easy along the way, so don't get frustrated and always be willing to ask for help, or to help others who ask you for help.

CHAPTER 12 SUMMARY

- Nanoscale materials can be used in many consumer products.
- Display technology has moved away from cathode ray tubes, and now takes advantage of nanotechnology like quantum dots, organic LEDs, polymer LEDs, and electronic ink.
- The textile industry implements nanoscale materials in products such as water-proof and stain-proof clothing and antimicrobial fabrics.
- There is a natural fit for nanotechnology and sports, since nanomaterials can make items lighter and stronger. Sports such as golf, tennis, cycling and skiing have benefited from nanotechnology.
- There are many promising applications of nanotechnology in every part of automobiles, from engine coolants to window coatings.
- Nanotechnology is used in the cosmetics industry in sunscreens and moisturizing anti-wrinkle agents.
- Microelectromechanical systems (MEMS) are being integrated into many home appliances like dishwashers with soil-sensors, and refrigerators with temperature sensors.
- Dog beds are being manufactured with nanoparticles to help prevent the spread of bacteria and odor from the fabric.
- Nanoscale materials may be useful in new types of packaging to keep food from spoiling.

REFERENCES

1. E. Boysen, N. Muir, *Nanotechnology for Dummies, 2nd Ed.*, For Dummies, 2011.
2. NanoWerk, www.nanowerk.com (June 29, 2015).

CHAPTER 13   NANOTECHNOLOGY COMPANIES

INTRODUCTION

In previous chapters of this textbook, you learned about the many ways in which nanotechnology can be used to improve everyday life. Nanotechnology is a large field that encompasses nanoscale materials such as individual nanoparticles like quantum dots and carbon nanotubes. There are companies that manufacture nanoscale materials, and also companies that use those materials and incorporate them into new products. Under the umbrella of nanotechnology are also the large fields of microelectronics and semiconductor manufacturing, the companies that create instruments for processing semiconductors, and equipment for visualizing small features and characterizing materials we cannot see by eye. The following chapter will outline a small selection of companies that are at the forefront of nanotechnology, nanoscale electronics, and nanoscale materials. While reading some of these brief descriptions, you may get a feel for the different areas of nanotechnology and which seems of greatest interest to you.

13.1   NANOSCALE MATERIALS

- Nanocomp Technologies, Inc.

  Nanocomp Technologies specializes in the production of carbon nanotubes. They commercially produce carbon nanotube fibers that get made into sheets, tapes, wires and yarns. These fibers are strong, lightweight, and conductive. As such, there are many applications for them, particularly within the energy, automotive, defense, and aviation industries. Nanocomp Technologies is located in Merrimack, New Hampshire and employs between 50 and 200 people. *www.nanocomptech.com*

- Evident Technologies

  Evident Technologies is a nanotechnology company based in Troy, New York. They commercially produce of quantum dots (semiconductor nanoparticles). Evident specializes in thermoelectronics (TE), technology in which electricity and temperature gradient can be converted from one to the other. TE are ideal for cooling small areas, or generating power by conversion of thermal energy into electricity. Additionally, Evident Technologies has partnered with electronics giant Samsung in the manufacture of quantum dot light emitting diode display technology. *https://**evident**thermo.com*

- NanoComposix

  NanoComposix is a company located in San Diego, California. They make and characterize any number of nanoparticles, specific to their customer's needs.

NanoComposix will customize silver, gold, silica, or iron oxide nanoparticles, for example, of whatever size, shape, and surface functionalization the customer requires for their applications. The nanoparticle systems are high quality and incredibly well-characterized.

www.nanocomposix.com

- Nanograde

    Nanograde also produces tailor-made nanoparticle systems specific to their customer's needs. The company is located in Switzerland and was founded in 2008. They specialize in nanoscale inks, suspensions of different nanoparticles in various solvents, nanoparticle dispersions, and colloidal solutions.

    www.nanograde.com

13.2 NANOTECHNOLOGY PRODUCTS

- 3M

    3M is a large company that employs over 80,000 people worldwide. They produce a wide range of products such as adhesives, dental products, car care products, optical films, abrasives, laminates, and electronic materials. The company has started incorporating nanoscale materials into a number of their products. For example, they have designed dental fillings, also known as dental restoratives, made with nanoscale materials to keep them stronger and brighter for longer lengths of time. Additionally, nanoscale materials are brought into their pharmaceuticals, adhesives, and energy departments.

    www.3m.com

- General Electric

    General Electric, or GE, is an American company that has locations all over the world. They have many segments aside from Technology, including Capital Finance, and Consumer & Industrial divisions. The history of the company is very rich, dating back to Thomas Edison's involvement in the late 1800s. As part of their research and development endeavors, GE has entered into the world of nanotechnology, always striving to develop innovative products and better materials. GE research has used nanotechnology for a variety of applications, including fuel efficiency improvement, faster electronic devices, and medical diagnostics, to name a few.

    www.ge.com

- SuperPower, Inc.

    SuperPower Inc. is a subsidiary of Furukawa Electronic Company that specializes in high temperature superconducting wires. Their superconducting wires are highly advantageous compared with traditional conductors such as copper. They are extremely efficient, and transport current with nearly no resistance to flow. According to their website, more than 7% of the energy currently generated in the United States is lost during transmission and distribution. It is estimated that half of this energy loss could be eliminated using high temperature superconducting wires. This would save billions of dollars and reduce the need for as many fossil fuels and greenhouse gas emissions.

SuperPower's wires are created via layer-by-layer thin film deposition techniques. Manufacturing is performed in Schenectady, New York, and research is done is Texas.

www.superpower-inc.com

- Toshiba

  Toshiba was founded in the late 1930s as what was known as Tokyo Shibaura Electronic K.K. Its name was officially shortened to Toshiba in 1978. As of 2010, Toshiba was the world's fifth largest vendor of personal computers, and fourth largest manufacturer of semiconductors devices. The company sells any number of electronics products, including Blu-ray players, air conditioners, air-traffic and railway control systems, medical equipment such as MRI scanners, personal computers, displays, and televisions.

  www.toshiba.com

Some other companies that include nanotechnology in their products are giants such as

- Nikon (*www.nikon.com*)
- Samsung (*www.samsung.com*)
- Intel (*www.intel.com*)
- IBM (*www.ibm.com*)

13.3 NANOSCALE ELECTRONICS AND EQUIPMENT

- SEMATECH (<u>S</u>emiconductor <u>M</u>anufacturing <u>T</u>echnology)

  SEMATECH is a consortium, or partnership, of various sectors within the electronics industry. Research and development groups, equipment and materials suppliers, universities, and government partners all cooperate via SEMATECH to reduce the amount of time it takes from innovation to manufacture of computer chips. SEMATECH began operating in the late 1980s to help solve common semiconductor manufacturing problems and to bring the United States to the forefront of semiconductor manufacturing on the world stage. Currently, SEMATECH has its headquarters in Albany, New York, and also has a development fabrication plant in Austin, Texas.

  Public.sematech.org

- GlobalFoundries

  GlobalFoundries is one of the world's largest manufacturers of semiconductor devices. Its headquarters are located in Milpitas, California, and it has fabrication plants in Singapore, Germany, and Malta, New York. GlobalFoundries employs approximately 10,000 people worldwide, and was created via a merger of the manufacturing side of Advanced Micro Devices (AMD) and Chartered Semiconductor in 2010. *www.globalfoundries.com*

- Taiwan Semiconductor Manufacturing Company Limited (TSMC)

  TSMC is the largest semiconductor manufacturing company in the world. TSMC makes chips for a number of high-tech companies all over the world. The company has wafer processing facilities capable of producing 200 and 300 mm wafers, and plans to build a 450 mm fab in the near future. *www.tsmc.com*

- Intel

  Intel is an American-based semiconductor chip maker with its headquarters in Santa Clara, California. They are the central inventors of microprocessor devices for personal computers, and also make devices including memory, graphics chips, motherboards, and network controllers. The company was founded by some of the first pioneers in semiconductor technology, Gordon Moore and Robert Noyce. *www.intel.com*

- Applied Materials

  Applied Materials is a company that provides equipment, along with software and maintenance services, to semiconductor manufacturing industries. They are also involved in flat panel display technology, solar photovoltaics, and other industries that use nanofabrication techniques such as patterning, formation of transistors, and interconnections.

  New York Times columnist Thomas Friedman quoted that, "Applied Materials is one of the most important companies you've probably never heard of," since most every commercially available electronic device contains some components that were made in part with Applied Materials equipment.

  Applied Materials has its main headquarters in Santa Clara, California. They also have locations throughout the United States and worldwide in Asia, Europe and the Middle East. *www.appliedmaterials.com*

- Tokyo Electron

Tokyo Electron Limited, or TEL, like Applied Materials, is a company that supplies equipment to semiconductor manufacturers to create chips for electronic devices. TEL creates equipment for any number of processes used in nanofabrication, for example thermal processing, photoresist coating, photoresist developing, chemical vapor deposition (CVD), doping, and reactive ion etching. As the name would suggest, TEL has its headquarters in Tokyo, but serves semiconductor manufacturing companies worldwide, throughout North America, Europe and Asia.
www.tel.com

- Screen Holdings Co., Ltd. (formerly Dainippon Screen or DNS)
  DNS produces a wide range of semiconductor manufacturing equipment. They produce wet stations for both 200 and 300 mm silicon wafer technology. DNS also manufactures annealing systems, as well as measurement and inspection systems for semiconductor companies.
  www.screen.co.jp

13.4   ANALYSIS AND CHARACTERIZATION

- FEI
  FEI is a world leader in the field of electron microscopes. They make some of the highest quality electron microscopes, both scanning electron and transmission electron, and also ion-beam microscopes. Their DualBeams™ microscopes combine both scanning electron microscopy with a focused ion beam microscope. FEI has about 1800 employees and serves more than 50 different countries around the world.
  www.fei.com

- JEOL\
  JEOL, originally Japan Electron Optics Laboratory Company, also specialized in the manufacture of high quality electron microscopes. Aside from transmission and scanning electron microscopes, JEOL manufactures other metrology equipment such as photoelectron spectrometers, X-ray fluorescence spectrometers, nuclear magnetic resonance, electron spin resonance, and mass spectrometers. JEOL employs nearly 3,000 people thought the world and has its headquarters in Tokyo, Japan.
  www.jeolusa.com

- Rigaku
  Rigaku is best known for its innovations in the field of x-ray spectroscopy. They make instruments for x-ray diffraction, which gives information about

crystalline structure. Their instruments can be used for thin film analysis, which is a very important part of nanofabrication. Rigaku instruments can be found in both industrial and academic settings. The company has approximately 1,100 employees in Asia, Europe, and the United States.
www.rigaku.com

- Lam Research Corp., and KLA Tencor
KLA Tencor Corporation, recently purchased by Lam Research, is an American company that manufactures control and yield management products for the nanoscale electronics industry. Their software and products help semiconductor fabs to manage yield when producing silicon chips. Lam Research Corporation is a leading supplier of wafer fabrication equipment and services.
www.lamresearch.com

NANOSPEAK - WITH MEGAN MOCK (CHAPTER 13)

1. Where do you work and what is your job title?
I work at CNSE (the College of Nanoscale Science and Engineering) for the Research Foundation of the State University of New York. I am a "Research Technician level 3" with a 'promotion' to Research Support Specialist in the works. I work in Lithography and will soon be working on the newest EUV (Extreme Ultraviolet) light lithography tool that a whole new addition to CNSE was built for! This tool is one of the first in the world and is very exciting new technology.

2. What is your educational background?
I went into a Nanotechnology associate's degree program from the field of welding.

3. How did you get interested in nanotechnology/where did you hear about it?
I actually endured a medical ailment that no longer allowed me to be a welder. So, as I have always enjoyed challenges and learning, I picked nanotechnology as the field to pursue for my future.

4. What advice would you give to someone who might want to get into the field?
Nanotechnology is growing at a fast pace and is needed for our technological future. I will always be able to grow, learn, and expand my future in the science and research field. Try to learn the history of when nanotechnology got started all the way to the newest pieces of equipment that are being produced.

CHAPTER 14   Careers IN NANOTECHNOLOGY

INTRODUCTION

Under the broad umbrella of nanotechnology, there are a number of different industries with various positions available. The following chapter will focus on the semiconductor manufacturing industry and wafer processing along with those companies that supply equipment to the fabrication plants. Careers in this field are in high demand and require some working knowledge of nanotechnology, science, and math, as well as strong interpersonal skills, communication skills, computer literacy, and since so much money is invested in the equipment and products, a person working in the semiconductor industry must be very responsible, punctual, and be able to work well in a team-oriented environment.

Be aware different companies may have their own particular position nomenclature, and the following is not necessarily completely universal.

14.1   MODULES/BAYS

Inside the semiconductor manufacturing fabrication plant ("fab"), there are a number of "Modules" or "Bays" wherein the various steps of the integrated circuit (IC) process take place. Each module requires its own set of experienced workers. Some of these areas within the fab are chemical mechanical polishing, lithography, cleaning ("wets" or "cleans"), etching, deposition/thin films or "producers," implant or diffusion, process integration, process integration yield enhancement, and metrology.

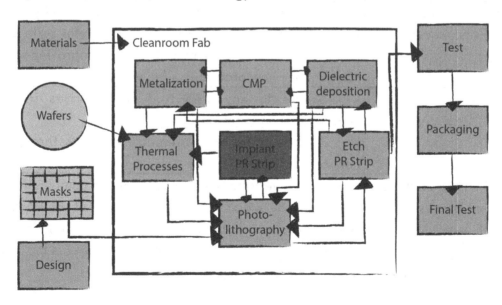

FIGURE 14.1

Figure 14.1 IC Fabrication Modules/Bays

Module/Bay	Description of Processes
Chemical Mechanical Planarization ("CMP")	-Removal process -Strips part of a deposited film by combining a chemical reaction with mechanical polishing techniques -Can also be used to remove bulk dielectric material or metal
Lithography ("Litho")	-Patterning process that transfers circuitry design from a mask or reticle onto the silicon wafer's surface
Cleaning ("Wets" or "Cleans")	-Where wet etching takes place -Chemical solutions used to dissolve materials on a silicon wafer's surface -Widely used for film stripping
Etching	-Reactive ion or dry etching -"Dry" etch process that uses plasmas to remove materials from a wafer's surface
Deposition/Thin Films ("Producers")	-Functional films are grown or deposited onto the silicon wafer's surface
Implant/Diffusion	-Dopant ions are forcefully added to a wafer in order to control conductivity
Process Integration ("PI")	-It requires hundreds of steps to finish an IC chip fabrication. -PI coordinates best route for wafers in order to make fabrication most efficient
Process Integration Yield Enhancement	-Yield refers to the functionality and reliability of integrated circuits produced on the wafer surface. -Yield engineers are responsible for improving yield which determines how profitable a fab will be.
Metrology	-Refers to any step that characterizes features via microscopy or spectroscopy. -Metrology (or measurement) happens throughout the entire fabrication process.

Within the fab there are workers at every module who function in a number of capacities. There are process workers, who help develop the actual processing of the silicon wafers, and well as equipment workers who focus on the individual tools used for processing. In general, due to the large number of steps required for wafer processing, a fab will run for 24 hours a day, 7 days a week.

14.2   THE CLEANROOM

All manufacturing is performed in a cleanroom environment in order to reduce any exposure to particulates. A cleanroom is a man-made environment with very low air particle counts.

Even the smallest of particles can destroy the functionality of an IC device. As feature size on devices is shrinking, the size of particles that will destroy a chip also decreases. As such, higher grades of air purity within cleanrooms are required. There are various classes of cleanroom, defined as having fewer than 'x' number of particles with diameter larger than 'y' per cubic foot of air. For example, a class 100 cleanroom is defined as having fewer than 100 particles with diameter larger than 0.5 ·m per cubic foot.

Inside a cleanroom, the floor is generally raised with grid panels. This allows the air to flow vertically from the ceiling to the area below the workspace. Air that comes into the cleanroom from outside is initially filtered to remove any dust particulates, and the recirculating air is passed through a high-efficiency particulate air ("HEPA") filter to remove any particles that might have been generated internally.

People working in the fab are the largest source of particle contamination. Therefore, very rigid cleanroom protocol must be followed by all those who work within the clean manufacturing environment. Staff members have to wear specially designed protective garments, often referred to as "bunny suits." The bunny suit is a coverall that includes a cap, booties, facemask, glasses and gloves. Workers enter and exit through airlocks, and cleanrooms are usually kept at positive pressure so if there are any air leaks, the air will move out of the room instead of allowing air to come in.

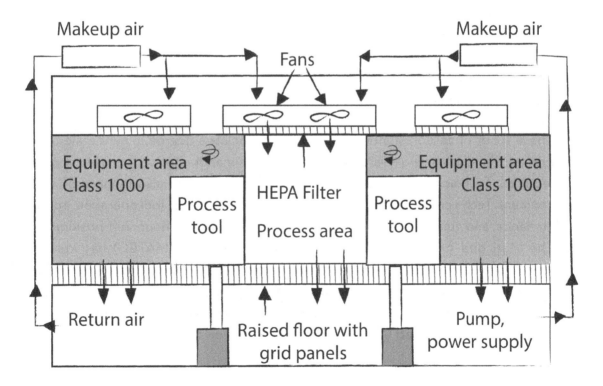

FIGURE 14.2

Figure 14.2 Cleanroom schematic

14.3 POSITIONS

The most entry-level positions can have many names such as "Process Operator," "Workstation Operator," or "Wafer Fab Technician," for example. These positions generally require an associate's degree in nanoscale materials or a related field. The Process Operator works in the cleanroom, typically for 12 hour shifts. A normal schedule would be 7:00 AM – 7:30 PM Sunday, Monday, Tuesday, and alternating Wednesdays, for example.

An operator would be responsible for operating equipment in any of semiconductor process modules or bays previously listed. They would have to perform such tasks as writing and updating operating procedures as needed, performing inspections, tests, minimal repairs, troubleshooting and operation of several process tools. They might be involved in new tool installation, troubleshooting, and qualification. The operator provides technical assistance to operations management and resolves tool problems. An understanding of computers is a must, as the tools are all generally run with computer software. The operator reports to their 'team lead,' who keeps track of all the tools in their area.

The position above an operator in the fab would be a technician. There are various levels of technician, for example

- Process technician;
- Associate technician;
- Senior technician;
- Principal technician;
- Staff technician;
- Maintenance technician.

These types of careers generally require an associate's degree in nanoscale materials or a related field, plus two years of semiconductor manufacturing experience may also be required. Technicians work both within the cleanroom and also in more of an office environment. Technicians perform any number of duties, including tool operation, equipment maintenance, and guidance of operators. These are also twelve hour shift positions.

The Maricopa Advanced Technology Education Center (MATEC) has identified a number of semiconductor manufacturing technician skill standards to address the skills that would be required of entry-level technicians on the job. The following table outlines some of the skill standards, along with conditions and behaviors expected of technicians. Most skills would be learned within the first six months on the job.

Standard #	Description	Examples of Condition/Behaviors
1	Implementing quality principles	-Interpret data of statistical process control charts -Communicate trends of machine performance -Analyze flow charts -Participate in teamwork and use problem solving techniques
2	Demonstrating working knowledge of basic electronic principles	-Measure voltage, current, and resistance using a multimeter -Analyze series and parallel circuits schematics containing resistors, inductors, and capacitors -Analyze digital and analog circuits
3	Operating equipment	-Adjust, calibrate, and test manufacturing equipment -Identify equipment inaccuracy -Troubleshoot manufacturing equipment
4	Processing Wafers	-Adhere to wafer handling and manufacturing procedures -Recognize the steps in the wafer manufacturing process -Recognize the function of process equipment
5	Troubleshooting and Repairing Electrical/ Electronic Systems	-Recognize electrical/electronic malfunctions -Conduct routine preventative maintenance using basic hand tools and replacement parts kits -Interpret electrical/electronic diagrams
6	Troubleshooting and Repairing Pneumatic Systems	-Recognize pneumatic malfunction and demonstrate cleanroom-approved workmanship skills when making repairs.
7	Troubleshooting and Repairing Hydraulic Systems	-Given a hydraulic pump, gauge, filter, accumulator, flow control valve, servo valve, directional valve, pressure control valve, preventive maintenance and minimal assistance, demonstrate cleanroom-approved workmanship skills when performing preventive maintenance on equipment
8	Troubleshooting and Repairing Mechanical/ Electromechanical Systems	-Given appropriate instructions, mechanical drive, servo, pump, stepper, clutch and speed reducer, perform mechanical and electromechanical diagnostic tests, perform preventive maintenance, and install mechanical and electromechanical components.

Standard #	Description	Examples of Condition/Behaviors
9	Troubleshooting and Repairing Vacuum Systems	-Be able to define vacuum terms -Describe operation of vacuum pumps and gauges -Install vacuum pumps and gauges according to manufacturer's specifications -Identify vacuum components and conduct vacuum diagnosis
10	Troubleshooting and Repairing Radio Frequency (RF) Systems	-Recognize the functions of the following RF subsystems: oscillators, amplifiers, filters, and coupling circuits -Identify RF connections and cabling -Demonstrate safety practices when working around an RF system
11	Operating Remote Systems	-Given a block diagram of a liquid heat exchanger, describe its function -Recall the reason for using deionized water in semiconductor manufacturing -Identify cause and effects problems with the gas delivery system
12	Performing Preventative and Routine Maintenance	-Use drills and attachments in accordance with cleanroom-approved techniques -Maintain a power drill -Fill out maintenance records including appropriate instructions -Given a service bulletin and an equipment training device, comply with requirements of the service bulletin -Interpret mechanical, electronic and electronic drawings -Demonstrate cleanroom-approved workmanship when using handtools such as screwdrivers, wrenches, sockets, hammers, pliers, wire strippers, torque wrenches, vises, etc.
13	Maintaining Automated Systems	-Calibrate robot coordinate systems -Troubleshoot, operate and maintain automated systems including robots

Standard #	Description	Examples of Condition/Behaviors
14	Implementing Manufacturing Technology and Techniques	-Conform to cleanroom protocol -Maintain chemical and gas delivery and disposal system
15	Utilizing Computers	-Read and respond to screen commands -Extract data according to specifications -Demonstrate specified skills requiring stands business functions of word processing and spreadsheets -Install and set up software -Given a microcomputer system, software, programming instructions and minimal assistance, design, debug, and run
16	Adhere to Basic Safety Principles	-Identify potential hazards and follow specified safety practices -Apply appropriate OSHA standards to a given situation -Explain common terms found on safety data sheets (SDS) -Describe the purpose and list common responsibilities of emergency response team
17	Applying Scientific Fundamentals	-Given a list of balanced chemical reactions, and a list of descriptions of processes that may include a chemical name, match the reaction to a specific semiconductor process -Explain fundamental behavior of acids, bases, and solvents -Write descriptions of the chemical interactions that take place during processing
18	Performing Mathematical Calculations	-Perform calculations and express results in metric units -Identify basic statistical methods -Given the necessary background theory of physics, chemistry, and electronics relevant to semiconductor technician competencies, perform calculations without error

Standard #	Description	Examples of Condition/Behaviors
19	Recognizing Workplace Fundamental Principles	-Given a list of activities in the workplace, tell which behaviors add to the cost of manufacturing a product and which ones do not -Describe the cost and value of employee benefits -Recognize ethical and non-ethical business practices -Write a personal goal from a technician's position that compliments the business' goal
20	Use Information Skills	-Read and perform the required actions in a procedure without errors -Perform documentation updates without errors -Schedule tasks in order of business priority -Deliver instructions enabling a peer to successfully complete a task
21	Employing Interpersonal Skills	-Demonstrate productive interpersonal relationships -Given specifications for a complex, team-oriented project, appropriate equipment and handtools, a list of team members and minimal assistance, demonstrate appropriate teamwork skills that result in effective work results -Describe constructive actions employees could take to contribute to change
22	Displaying Appropriate Personal Qualities	-Given scenarios in which one's self esteem in challenged, demonstrate appropriate responses -Given a policy, a set of scenarios concerning harassment, discrimination, etc., demonstrate appropriate social skills -Given a scenario in which you have a challenging amount of work to accomplish in a minimal amount of time, describe strategies to accomplish all work assigned

Engineers in a fab are generally those employees with a bachelor's degree and upward. They are responsible for the individual tools, as well as for the creation of recipes for running the tools. Unlike operators and technicians, engineers generally do not work twelve hour shifts 3-4 days a week, but follow a more traditional 5 day a week schedule. The scientists that design the actual circuitry pattern that gets printed onto IC chips generally requires a PhD. This specialized design process is done outside of the fabrication area by each individual company requiring chips be manufactured for their products.

NANOSPEAK – WITH DYLAN CLARK (CHAPTER 14)

1. Where do you work and what is your job title?
    Work station operator at CNSE.

2. What is your educational background?
    A high school diploma, an associate's degree in the humanities and liberal arts, and an associate's degree in nanoscale materials.

3. How did you get interested in nanotechnology/where did you hear about it?
    My interest in the field of nanotechnology was initially sparked by talking to students of the nanoscale materials program while I was working on my humanities degree. They told me about all of the educational and job opportunities in nanotechnology and it seemed like a great fit for me.

4. What advice would you give to someone who might want to get into the field?
    A community college education in nanotechnology is promising and rewarding, and has afforded me an internship with CNSE and GlobalFoundries. There are many potential career choices for you if you pursue this type of path.

REFERENCES:

1. M. Quirk, J. Serda, *Semiconductor Manufacturing Technology*, Prentice Hall, 2000.
2. Maricopa Advanced Technology Education Center, matec.org (July 15, 2015).

CHAPTER 15   Health and Safety

15.1   POTENTIAL HEALTH RISKS

In recent years, research on the potential health risks involved in working with nanoscale materials has increased dramatically. Use of engineered nanoscale materials has become very widespread, and therefore the potential for exposure to these types of particles has also increased. It is still not completely clear how such small particles interact with the human body. Possible exposure routes of nanoparticles include, but are not limited to inhalation (though the respiratory tract), dermal (thorough the skin) and oral (through the mouth).

There is no doubt that the exceptionally small size of nanoparticles makes them very useful. It has been shown that nanoparticles have very different properties than bulk materials, and therefore have an unlimited number of potential new uses. It is these same novel properties, such as increased surface area and chemical reactivity however that make nanomaterials potentially dangerous to humans.

One such example is graphite, or carbon sheets. Graphite's properties are very well known and well-characterized. There are particular guidelines that relate to handling graphite based on toxicological data. Fullerenes, like carbon nanotubes and buckyballs, are legally categorized as graphite. Carbon nanostructures do not behave in the same way as bulk forms of carbon (Chapter 5). Studies as far back as nine and ten years ago have shown that fish who were exposed to buckyballs, even at low concentration, experienced extensive amount of brain damage. This would indicate that such small particles have the ability to cross the blood-brain barrier. In addition to brain damage, the fish also experienced liver damage, which affects their entire physical and chemical bodily processes. While studies cannot be conducted directly on humans, it is clear that accumulation of small particles could potentially have concerning health effects.

Other studies involving needle-like carbon nanotubes have shown that they may accumulate in the lungs upon inhalation. This is compared to the danger associated with asbestos dust. In the late 19th century, asbestos became very popular because it has some very desirable properties such as high tensile strength and resistance to fire, chemical damage, and electricity. Its structure, however, contains thin fibrous crystals, which have now been proven to cause serious illness upon inhalation, such as lung cancer and mesothelioma. Even though asbestos has not been used since the 1970's, nearly 3,000 deaths per year can be attributed to exposure decades past.

In addition to fullerenes, other types of nanoparticles have been studied and shown to have adverse effects on humans. A 2002 study of cadmium selenide nanoparticles, or quantum dots, showed that they can cause cadmium poisoning in humans. Cadmium is very toxic, and can case many adverse symptoms, including respiratory tract and kidney problems. If cadmium is ingested, the liver can become immediately damaged, and the kidneys may lose their function. This damage is irreversible.

Many commercially available sunscreens contain titanium and zinc nanoparticles (Chapter 12). As far back as 1997, it has been shown that these nanoparticles can cause free radicals to form that are damaging to DNA. Damaged DNA may lead to cancer or birth defects. Newer studies have suggested that nanoparticles may cause damage to our genetic material without even having to cross cell membranes into cells. It is now thought that we may need to rethink what the term 'exposure' means when talking about nanoparticles.

While it is universally agreed upon that nanoscale materials are interesting and useful, more research is needed to determine the possible dangers associated with these newly engineered materials. While our bodies have natural immune defenses for natural foreign bodies such as viruses, nanotechnology is introducing entirely new types of substances to which we do not yet have immunity. In general, the smaller the nanomaterials are, the more toxic and bioactive they may be. As size decreases, ability to interact with skin, lungs, blood, brain, etc. increases.

Even in semiconductor manufacturing where individual nanoparticles are not synthesized, there are a large number of hazardous and toxic compounds used on a regular basis. Potential health effects from these chemicals are available via documents known as Safety Data Sheets (SDS). A SDS will give personnel the proper and safe handling procedures for working in a safe manner with each compound. Information given includes physical data such as melting point, boiling point and flash point, health effects, first aid procedures, chemical reactivity, proper storage and disposal, which protective equipment to use, and spill-handling procedures.

Chemical Compounds	Use In Manufacturing	Hazards
Oxygen gas and hydrogen gas	Oxidation of silicon	Fire or explosion hazard at high temperatures
Ethyl lactate n-butyl acetate Xylene 2-pentanone Propylene glycol Ethyl ether	Photoresist solvents	-Highly flammable -Severe eye irritation -Respiratory irritation -Potential for organ damage if exposure occurs
Sulfuric acid Hydrofluoric acid Phosphoric acid Acetic acid Nitric acid	Wet etching	-Highly corrosive -Pulmonary edema -Eye irritation -Burns skin

Chemical Compounds	Use In Manufacturing	Hazards
Arsine Arsenic pentafluoride Diborane Boron trifluoride Phosphine Phosphorus pentafluoride	Dopants	-Corrosive -Pulmonary edema -Affects the central nervous system -Kidney failure -Hemolysis
Silane Tungsten hexafluoride Dichlorosilane Ammonia Ozone Silicon tetrachloride Tetraethyl orthosilicate	Thin Film Deposition	-Asphyxiation -Pulmonary edema -Airway irritation -Hemolysis -Skin and mucous membrane irriation

15.2 MANAGEMENT

Occupational Health and Safety, PPE

The field of nanotechnology is relatively new and therefore there is a general lack of information as to the risks they pose to humans and how to best control those risks. While limited to animal studies, it does appear that inhalation, skin contact, and ingestion of nanoparticles can cause adverse health effects. So far, it appears inhalation may pose the greatest risk.

Animal studies have shown that nanoparticles are more likely to accumulate in the respiratory tract than their bulk counterparts. Again, research has shown that small nanoscale particles may cross cell membranes and be more toxic and biologically active due to their large surface area. Additionally, since they have negligible mass, nanomaterials can stay airborne for longer periods of time than bulk materials.

Perhaps the applications of nanotechnology that pose the greatest health risks are those that include generation of nanoparticles in the gaseous phase or aerosolization of nanoparticles. Using nanoparticles as powders, cutting, grinding, pouring, and mixing also pose higher inhalation risks.

When a person is working with nanoparticles, they should be aware of any standard operating procedures (SOP) that exist that that particular site. They should minimize the potential for inhalation exposure and skin contact, for example by working in a fume hood. Good hygiene such as consistent handwashing should always be practiced. Additionally, if there are instructions from the nanoparticle vendor or manufacturer, they should be carefully followed. Nanoparticles should always be handled, stored, and transported in a closed, sealed container to prevent them from escaping into the air, and wet methods should be used to clean up work areas to keep nanoparticles into a solution and not into the air.

Personal protective equipment, or PPE, should always be used when handling nanoparticles. Polymer gloves, like nitrile, should always be used, and even wearing two pairs so as to prevent the small particles from penetrating the gloves into the skin. The effectiveness of various face respirators are still being studied. When determining which respirator is most appropriate to use, the type and size of the nanomaterials being used should be evaluated and researched. Common dust masks, for example, are not effective for use when working with nanoparticles. Additionally, chemical splash goggles and lab coats are to be used when working with nanoscale materials. The lab coats need to be laundered on a regular basis, and are never to be brought out of laboratory areas. Closed-toe shoes are recommended, as well as covering them with booties and covering arms with sleeve covers to reduce any risk of skin contact.

There should always be a written plan in place for any spills of nanoparticle solutions and any clean up required for nanoscale materials. In general, gloves should always be worn for cleanup. To avoid making nanoparticles airborne, practices such as sweeping should be avoided. Instead, wet wiping methods are best, wherein a liquid material is placed down to dissolve the particles into a solution, followed by an absorbent material to pick up the suspension. Any absorbent materials such as towels should never be used again, and should be placed in a sealed plastic bag, followed by a second sealed bag. This outer bag should be carefully labeled with the name of the nanomaterials.

Waste that contains nanoparticles should only be handled in a well-ventilated fume hood. Any contaminated materials such as gloves or bench papers should be double bagged and labeled well to indicate not only the material but 'nano' material.

15.3   REGULATIONS

LEGAL

A legal framework for nanotechnology began under the administration of President Clinton with the signing of the National Nanotechnology Initiative (NNI). The NNI is meant to serve as a point of collaboration for Federal agencies involved in nanotechnology research. The initiative was written with the following four goals in mind:

1. Advance a world-class nanotechnology research and development program
2. Foster the transfer of new technologies into products for commercial and public benefit
3. Develop and sustain educational resources, a skilled workforce, and the supporting infrastructure and tools to advance nanotechnology
4. Support responsible development of nanotechnology

Researchers are still trying to determine how to carry out nanotechnology research and development in a "responsible" manner.

Currently, the National Institute for Occupational Safety and Health (NIOSH), a sector of the Centers for Disease Control and Prevention (CDC), is the main federal agency conducting research and giving guidance on occupational health and safety implications of working with nanoscale materials. The research they conduct seeks to answer the following questions:

1. How might workers be exposed to nanoparticles in the manufacture or industrial use of nanomaterials?
2. How do nanoparticles interact with the human body's systems?
3. What effects might nanoparticles have on the body's sysyems?

NIOSH has identified ten different critical areas with will help to address current knowledge gaps, develop strategies, and provide recommendations to those working with nanoscale materials. The CDC website provides descriptions for how the NIOSH is conducting research in these critical areas, and potential implications for the workplace.

1. Toxicity and Internal Dose

The NIOSH investigating what are the physical and chemical properties of nanoscale materials that influence their toxicity to humans. It is important to understand both the immediate and long term effects on the body that exposure to these materials may have. In addition, the NIOSH is researching how nanoparticles interact with and disrupt our natural biological mechanisms. It is important to create a standard model that will help to assess the potential hazards of working with engineered nanoscale materials.

2. Risk Assessment

NIOSH is committed to determining the likelihood that current exposure-response data could be used to identify and assess potential occupational hazards. Additionally, it is important to develop a framework for evaluating potential hazards and predicting the occupational risk of nanoparticle exposure.

3. Epidemiology and Surveillance

Epidemiology is the branch of medicine that deals with the incidence and prevalence of disease in large populations of people. NIOSH is currently evaluating existing epidemiological workplace studies where nanoscale materials are used and identifying new studies that might advance our current understanding of working with nanomaterials. It has also been determined that it is critical to integrate nanotechnology health and safety issues into existing hazard surveillance methods and determine if new screening methods are needed.

4. Engineering Controls and Personal Protective Equipment

NIOSH is dedicated to testing the effectiveness of engineering controls in reducing occupational exposures to nanoparticles and to developing new controls where needed. They also study current personal protective equipment and try to improve it. It was mentioned that we need to determine effectiveness of respirators and develop recommendations to workers to prevent or limit occupational exposures.

5. Measurement Methods

Methods of measuring which masses of nanoparticles can be inhaled into the lungs from the air are being evaluated, and then it is also being determined if this measurement can be applied to measure nanoparticle size. Practical field-testing methods are also being determined to quickly and accurately measure any airborne nanoparticles in the workplace.

6. Exposure Assessment

Key factors that influence the production, dispersion, accumulation, and re-entry of nanoscale materials into the workplace are being determined. Possible exposures when nanoparticles are inhaled or touched to the skin are being assessed, along with how possible exposures differ by workplace.

7. Fire and Explosion Safety

NIOSH is working to identify physical and chemical properties of materials that contribute to dustiness, combustibility, flammability, and conductivity of nanomaterials, along with recommending alternative work practices to eliminate or reduce workplace exposures to nanoparticles.

8. Recommendations and Guidelines

Until we know for certain what the effects of nanoparticle exposures are, NIOSH is using the best available science to make interim recommendations for workplace safety and health practices during the production and use of nanoscale materials. As more data are obtained, occupational exposure limits will be continuously updated.

9. Communication and Information

NIOSH is establishing partnerships to allow for companies and agencies to identify and share research needs, approaches, and results. They are also developing and disseminating training sessions and educational materials to workers in this new and developing field.

10. Applications

NIOSH is identifying uses of nanotechnology for application in occupational safety and health as well as evaluating and distributing effective applications to workers and occupational safety and health professionals.

NANOSPEAK – WITH MATTHEW PETRILLOSE (CHAPTER 15)

1. Where do you work and what is your job title?

   I work at the College for Nanoscale Science and Engineering as a photolithography technician II.

2. What is your educational background?

   I actually have four degrees! An associate's degree in hotel restaurant management, a bachelor's degree in secondary education in social studies, a master's degree in curriculum and instruction, and most recently received an associate's degree in nanoscale materials technology.

3. How did you get interested in nanotechnology/where did you hear about it?

   I got interested in nanotechnology when I started going back to school for electrical engineering. I was retaking calculus and my professor suggested I get into the field of nanotechnology.

4. What advice would you give to someone who might want to get into the field?

   Prepare yourself to be continually learning both on and off the job. Also, don't be afraid to fail. Sometimes that will teach you more than any class you'll ever take.

REFERENCES

1. M. Hull, D. Bowman, *Nanotechnology Environmental Health and Safety 2nd Ed.,* William Andrew, 2014.
2. National Nanotechnology Initiative, Nano.gov (July 30, 2015).
3. Centers for Disease Control and Prevention, the National Institute for Occupational Safety and Health, cdc.gov/niosh (August 5, 2015).

CONCLUDING REMARKS

It is my hope that through reading this book you have gotten a picture of the diversity and great potential of this broad field we call nanotechnology; a specific subset of materials science. There are two main divisions within the discipline, namely bottom-up nanotechnology, or individual nanoparticles, and top-down nanotechnology, where nanoscale features are built into bulk materials. There are endless possibilities for applications in all fields of science and technology. Sometimes even the smallest of technologies can lead to big, rewarding careers, and it is my goal to have opened your eyes to opportunities in this exciting field. There are many feasible pathways and ways to make a career in nanotechnology ranging from high school education all the way up to Ph.D. level studies. What I can say with certainty is that there is no chance nanotechnology will disappear from our society, and just by showing interest, you have put yourself in a place where you may reap benefits from a career in this high need, highly skilled discipline. And so I conclude by saying keep learning, keep asking questions, and keep imagining the possibilities for nanotechnology in our future. It is all in your hands!

GLOSSARY OF TERMS

A

Actuator: A type of motor used to manipulate or move components of an electrical device.

Advanced Material: Material made for 'high-tech' applications, for example semiconductors and nanoscale materials.

Aerogel: A colloidal suspension wherein particles of a gas are suspended in a continuous solid medium.

Allotropes: Various forms of an element bonded in different manners.

Alloy: Materials made of a combination of metals.

Amorphous: A material wherein the atoms are not ordered in any specific way.

Anion: A negatively charged species.

Anisotropic: A process that takes place only in one direction.

Annealing: In nanofabrication, a process that uses high temperature to restore the crystalline form of a surface.

Anode: Electrode into which electric current flows in electrical devices.

Antiferromagnetism: Phenomenon wherein all of a material's 'atom magnets' are pointing in opposite alternating directions and cancel out.

Antioxidant: A molecule that can stop the oxidation of other molecules that result from the production of free radicals.

Array Detection Technology: Nanoscale arrays of pads that can be constructed onto a single chip using standard nanofabrication techniques.

Atom: An extremely small particle of matter that maintains its identity during a chemical reaction.

Atomic Force Microscope (AFM): A scanning probe microscope that gives high resolution surface images down to the nanometer level.

Atomic Number: Total number of protons in an atom's nucleus.

B

Band Gap: An energy range that is between the valence and conduction bands. Energy range in a solid where no electronics can exist.

Battery: A device that converts chemical energy into electrical energy.

Binary System: Computer processing instructions represented using 0's and 1's.

Binder: See 'Matrix'

Bioavailability: The presence of a drug where it is needed and can be most effective.

Biocompatibility: Ability of molecules to flow through the blood stream without being attacked by the immune system.

Biomimicry: Any science wherein naturally occurring phenomena are used to solve human problems.

Biosensor: A device that responds to a specific chemical species within biological samples.

Bipolar Junction Transistor: Type of transistor that relies on the contact of two types of doped semiconductor material for operation as amplifiers or switches.

Bit: A single 1 or 0 within the binary system.

Bonding: The movement and/or sharing of electrons between atoms to achieve an octet.

Bottom-Up Nanofabrication: A process by which nanostructures are built up from individual atoms or molecules.

Brownian Motion: The random movement of particles suspended in a fluid resulting from their collisions.

Buckminsterfullerene: Spherical molecule made of 60 carbon atoms in a fused ring structure.

Bucky-ball: See *Buckminsterfullerene*.

Bulk Matter: Matter with the order of 6.023×10^{23} or more atoms.

Byte: A combination of 8 bits.

C

CIGS: Copper Indium Gallium diselenide semiconductor material.

Carbon Cycle: Natural cycle wherein plants, algae, and certain bacterial convert carbon dioxide into life-sustaining surgars.

Carbon Nanotube: An allotrope of carbon with a cylindrical structure.

Cancer Cells: Abnormally dividing and growing cells that proliferate in an uncontrolled way.

Capacitance: A material's ability to store electrical charge.

Capacitor: Device that holds electrical charge using at least 2 electrical conductors separated by an insulator.

Catalyst: A material that increases the rate of a reaction.

Cathode: Electrode OUT of which electric current flows in an electrical device.

Cathode Ray Tube: A vacuum tube with an electron beam source.

Cation: A positively charged species.

Cell Membrane: A double layer of phospholipids that help separate the inside of a cell from its surrounding environment.

Ceramic: A compound that occurs between a metallic or semimetallic element and a non-metallic element (usually oxygen, nitrogen or copper).

Ceramic Matrix Composite: A composite wherein a ceramic serves as the main component.

Chemical Reaction: The rearrangement of atoms in the reacting substance(s) to create new chemical combinations.

Chemical Vapor Deposition: A process by which a gaseous material undergoes a chemical reaction and is deposited as a solid film on a substrate

Chips: See *'Integrated Circuit Device'*

Composite Material: A material made up of two or more materials with different physical properties.

Compound: A type of matter composed of two or more elements chemically combined in fixed ratios.

Colloidal Gold: A suspension of micron-sized or smaller gold particles in a liquid.

Colloidal Nanoparticle Synthesis: A method of fabricating nanoparticles that uses salts suspended in a solution as a precursor material.

Colloidal Silica: A suspension of micron-sized or smaller silica (silicon dioxide) particles in a liquid.

Colloidal Silver: A suspension of micron-sized or smaller silver particles in a liquid.

Computer Processor: The central processing unit within a computer that carries out instructions of a computer program.

Concrete: A mixture of small, coarse particles embedded in cement.

Condensation: A phase transition wherein a material changes from the gaseous state to the liquid state.

Conductor (Electrical): An electrical conductor is a type of material which permits the flow of electrical charges in one or more directions with ease.

Constructive Interference: Two waves interacting that have the same wavelength and same phase resulting in double the amplitude.

Contrast Agent: A substance used to enhance the images obtained by medical imaging devices.

Covalent Bond: Sharing of electrons by atoms in order to achieve a full outer electron s hell.

Crystalline: A material that possesses a regular, ordered atomic structure.

D

Delocalized Electrons: Electrons in a molecule that are not associated with a single atom or covalent bond.

Deposition:

1. A phase transition wherein a material changes from the gaseous state to the solid state.

2. The step in nanofabrication wherein thin layers are formed.

Desalinization: Any process that aids in the removal of salt from water.

Destructive Interference: Two waves interacting that have the same wavelength and opposite phase resulting in no remaining wave.

Dielectric: *See 'Insulator'*

Diffraction Grating: A material containing small slits with sizes is on the order of the wavelength of visible light.

Diode: A two-terminal electronic component which only allows current to flow in one direction.

Dip Pen Lithography: Lithography process that uses an atomic force microscope probe tip to write patterns on a substrate.

Directional Bond: Bonds are in fixed positions in relation to the surrounding atoms.

Discrete Device: Any basic, single electrical component.

Dopant: Impurity atoms added to semiconductor materials.

Doping: Process of adding impurity atoms to a semiconductor material.

Dry Cell Battery: A battery wherein the electrolyte is immobilized in a paste form.

Ductility: The degree of plastic deformation that occurrs in a material at the point of fracture.

E

Elastic Deformation: Non-permanent change in a materials shape after an applied force is released.

Elastomer: A polymer that is very pliable, often called "rubber."

Electrical Conduction: Ease at which a material transmits an electrical current under an applied electric field.

Electrical Conductivity: A measure of a material's ability to conduct electric current; reciprocal of electrical resistivity.

Electrochemical Cell: A device powered by an oxidation-reduction reaction wherein one material loses electrons and another gains electrons.

Electrolysis of Water: The decomposition of water into hydrogen gas and oxygen gas by passing an electric current through.

Electrolyte: A compound that has been ionized.

Electron: Negatively charged particles that are constituents of all atoms.

Electron Beam Lithography: A lithography technique that uses a high energy beam of electrons to draw patterns directly into photoresist.

Electron Shell: Paths that electrons follow as they move around an atom's nucleus.

Electronegativity: The measure of the ability of an atom to draw electron density toward itself in a covalent bond.

Electronic Ink: Small black and white pigments that can move up and down in liquid when electrically charged to form an image.

Electronic Paper: A paper-thin flexible screen that can be written on or have information downloaded to.

Element: A type of matter composed of only one kind of atom.

Energy Bands: Description of energy ranges an electron within a solid material may have and may not have.

Engineering Strain: Deformation of a solid that results from engineering stress.

Engineering Stress: The ratio of applied force per cross sectional area of a material, often defined as force per area.

Equilibrium: The state when the rates of forward and reverse reactions or phase changes are equal.

Etching: In nanofabrication, any technique that removes materials from a substrate's surface.

Excited Electron: An electron with excess energy.

Extreme Ultraviolet Lithography: Lithography techniques that employ extreme ultraviolet electromagnetic radiation with wavelengths of 11-14 nanometers.

Extrinsic Semiconductor: A semiconductor whose electrical behavior is dictated by added impurity atoms.

F

Ferromagnetism: Phenomenon wherein a materials 'atom magnets' all point in the same direction resulting in an inherently magnetic material.

Free Radicals: Atoms or molecules with unpaired electrons.

Freezing: A phase transition wherein a material changes from the liquid state to the solid state.

Frequency: The number of wavelengths of a wave that pass through a fixed point in one second.

Fuel Cell: A device that generates electricity via a chemical reaction between oxygen and an oxidizing fuel source.

Functionalization: Chemical alteration of a surface.

G

Gas: The state of matter that is easily compressible, and has no fixed shape or volume.

Gecko Tape: A material made by mimicking the ability of a gecko to scale walls and hang from ceilings.

Generator: A device that converts mechanical energy into electrical energy.

Glass: A special type of ceramic with amorphous atomic structure.

Glioblastoma: A particularly aggressive form of brain tumor with finger-like projections.

Graphene: An allotrope of carbon wherein the atoms are arranged in a regular hexagonal pattern and is one atom thick.

Grätzel Cell: A type of solar cell that uses organic dyes.

Greenhouse Effect: A process by which thermal heat is absorbed by greenhouse gases and re-radiated in all directions.

Greenhouse Gas: Gases that contribute to trapping of warm air in the atmosphere.

Ground State: Lowest energy level of an electron.

Group: Elements in any vertical column on the periodic table.

Grow-and-Place: Method of fabricating nanoelectronic devices wherein individual nanoparticles are first grown and then placed on a substre.

Grow-in-Place: Method of fabricating nanoelectronic devices wherein individual nanoparticles are synthesized on the substrate where it will be used.

H

Hardness: Resistance of a material to scratching of its surface.

Heat Capacity: A material's ability to absorb heat from its external surroundings.

Highest Occupied Molecular Orbital: The molecular orbital or energy band with the highest energy that also contains electrons.

Hole: Concept that describes the lack of an electron at a position where one could exist.

Hybrid Nanofabrication: A process by which nanostructures are made by combining both top-down and bottom-up nanofabrication techniques.

Hybrid Orbital: The result of different orbitals from the same atom combining.

Hydrogen Bond: A secondary bond wherein a hydrogen atom covalently bonded to a highly electronegative atom (oxygen, nitrogen, fluorine) for an electrostatic link with another electronegative atom in another molecule.

I

Insulator (Electrical): An electrical insulator is a material whose structure does not allow electric charges to flow freely and therefore cannot conduct an electric current.

Integrated Circuit Device: A device that contains multiple electrical components connected to the same piece of semiconducting material often referred to as "chips."

Integrated Gasification Combined Cycle Technology: Clean coal technology that involves turning coal into a gas and removing the carbon dioxide before it gets burned.

Intrinsic Semiconductor: A semiconductor material whose electronic character is inherent in the material.

Ion: A charged atom or group of atoms.

Ion Concentration Polarization: A phenomenon that occurs when an ion current is passed through an ion-selective membrane.

Ion Implantation: In nanofabrication, a process that adds dopant atoms to silicon by forcefully bombarding the surface with a high energy beam of ions.

Ionic Bond: A bond between metal and non-metal atoms wherein the metal loses electrons (becomes a cation) and the non-metal accepts electrons (becomes an anion).

Isotope: Atoms in which the nuclei have the same atomic number but different atomic mass.

Isotropic: A process that occurs equally in all directions.

L

Langmuir-Blodgett Film: One or more monolayers of a carbon-based material deposited from the surface of a liquid onto a solid by immersion on the solid into the liquid.

Light-Emitting Diode (LED): Diode that emits visible colors of light.

Light Interference: Result of two waves interacting with one another.

Liposome: An artificially created phospholipid bilayer.

Liquid: The state of matter which is relatively incompressible, has a fixed volume, but has no fixed shape.

Liquid Crystal: A unique material that is pourable like a liquid but has regular crystalline structure like a solid.

Liquid Crystal Display: A flat panel display that uses light modulating properties of liquid crystals to generate images.

Lithography: In nanofabrication, a process wherein a pattern is transferred onto a substrate.

Lotus Effect: The very high superhydrophobicity exhibited by the leaves of a lotus flower.

Lowest Unoccupied Molecular Orbital (LUMO)/ Empty or Conduction Band: The molecular orbital or energy band with the lowest energy that does not contain electrons.

M

Magnetic Resonance Imaging (MRI): Medical imaging technique used to visualize internal body structures which cannot be seen via x-ray.

Magnetic Storage: Storage of digital data on a magnetized material.

Mass Number: Total number of protons and neutrons in an atom's nucleus.

Material: A substance out of which useful things can be made.

Material Modification: In nanofabrication, any process that is used to tailor physical properties of a surface.

Matrix or Binder: The bulk component in a composite material.

Melting: The phase transition wherein a solid state material changes to a liquid state material.

Membrane: A physical barrier that will allow certain substances to penetrate easier than others.

Mesoporous: A material with pores between 2 and 50 nanometers in size.

Metal: Elements such as gold and copper which appear on the left side of the periodic table and are often characterized by opacity, ductility, and conductivity.

Metal Oxide Semiconductor Field Effect Transistor (MOSFET): A device used for amplifying or switching electrical signals with a source, gate, and drain.

Metallic Bonding: The way in which atoms of metals are held together. In metallic bonding, free electrons hold together positively charged metal cations.

Metalloid: See 'Semimetal'

Metal Matrix Composite: A composite wherein a metal serves as the main component.

Microelectromechanical Systems (MEMS): Small devices made of components on the order of 1-100 micrometers.

Microcontact Printing: A lithography technique that uses a soft rubber stamp to transfer patterns to a substrate.

Microscopy: A process that employs microscopes to show size, shape and structure of nanomaterials.

Modulus of Elasticity: Mathematical description of a material's tendency to deform when a force is applied.

Moletronics: Electronic devices that are based on individual molecules.

Moore's Law: An observation made by Gordon Moore that over the history of computer processors, the number of transistors on a chip doubles approximately every 18 months.

N

N Type Semiconductor: An extrinsic semiconductor doped with an atom that has 5 valence electrons.

Nanocomposite: A multiphase material wherein one component has dimensions on the nanoscale.

Nanoelectronics: Electronic devices that are based on individual nanoparticles.

Nanoemulsion: *See 'Nanosphere'*

Nanofabric: Fabrics made of a composite of nanoscale materials and fabric threads.

Nanofabrication: The process by which nanostructures are made.

Nanofluid: Engineered materials that contain nanoparticles suspended in a liquid.

Nanogenerator: Self-powered and self-sustaining generators that use nanoscale materials.

Nano-imprinting: A lithography process that presses a mold with nanoscale features on it into resist to transfer a pattern into a substrate.

Nanometer: 1 billionth of a meter.

Nanosphere: Lipid-based nanoparticles that encapsulate therapeutic agents and can easily merge with cell membranes.

Nanotechnology: Man made technology or material that contain at least one dimension with size on the order of one nanometer.

Neutron: Nuclear particle that carries no charge.

Noble Gas: Elements in group VIII of the periodic table whose outermost electron shells are filled.

Non-Metal: An element on the right-hand side of the periodic table that does not exhibit the characteristics of a metal.

Nucleus: Positively charged center of an atom that contains protons and neutrons.

O

Octet: Condition wherein an atom has eight electrons in its outermost electron shell.

Optical Microscope: Microscope that employs light and a series of lenses to magnify the view of materials.

Orbital: Mathematical probability that describes a volume of space around an atom's nucleus that an electron is most likely to occupy.

Organic Light-Emitting Diode (OLED): Light emitting diode in which the electroluminescent layer is an organic film that emits light when an electric current is applied.

P

P-N Junction: The boundary between a P-Type and N-Type semiconductor material in a device.

P Type Semiconductor: An extrinsic semiconductor doped with an atom that has 3 valence electrons.

Pathogens: Disease-causing organisms.

Paramagnetism: Phenomenon wherein a material's 'atom magnets' are all pointing in random directions and cancel out.

Performance: How a material is used.

Period: Elements in any horizontal row on the periodic table.

Periodic Table: Arrangement of elements into rows and columns by increasing atomic number.

Phase Diagram: A graph that summarizes the conditions under which a material is solid, liquid, or gas.

Phonon: Level of vibrational energy in a solid material.

Phospholipid: A molecule containing two parts; one that is hydrophobic and one that is hydrophilic.

Photoelectric Effect: Emitted electrons generated from matter absorbing energy in the form of electromagnetic radiation.

Photolithography: Lithography process that uses photoresist and light to transfer patterns onto a substrate.

Photon: Particles of light that can be thought of as tiny packets of energy.

Photoresist: A light-sensitive polymer material used in lithography steps of nanofabrication.

Photosynthesis: Process wherein sunlight and carbon dioxide are converted into sugars.

Photovoltaic Cell: Device that converts light into electricity.

Physical Vapor Deposition: A process by which a material is vaporized by physical means and is deposited as a solid film on a substrate

Piezoelectric: Materials that generate an electric potential in response to an applied mechanical stress.

Pixel: Individual point in a display device.

Plasma Etching: In nanofabrication, a process that uses gaseous chemicals to remove material from a substrate's surface.

Plastic Steel: A composite plastic modeled after the lining of oysters, mussels or abalone shells.

Plastic Deformation: Irreversible deformation of a material after having undergone elastic deformation.

Point-of-Use: Act of purifying water where it will be consumed.

Polar Covalent Bond: Covalent bond wherein electrons are shared unequally.

Polymer: Long-chained molecules made up of repeating parts.

Power Density: The amount of power a battery outputs per unit volume, or weight of the battery.

Primary Battery: Disposable batteries.

Primary Bond: Bonds directly involved in holding atoms together.

Processing: How a material is treated in order to change its physical and/or chemical properties.

Properties: Both physical and chemical behavior of a material.

Proton: Nuclear particle that carries positive charge.

S

Sandwich Composite: A composite material wherein two thin yet sturdy skins are attached to a lightweight, thick core.

Scanning Electron Microscope: A microscope that traces a high energy beam across a conductive sample to provide magnified views of a material.

Scanning Probe Microscope: Any microscope that uses a physical probe to give information about the surface topography of a sample.

Scanning Tunneling Microscope: A type of scanning probe microscope that can produce images down to the atomic level.

Scientific Notation: A method of expressing numbers as the product of a decimal between 1 and 10 and an appropriate power of 10.

Pyroelectric: Materials that generate an electric potential upon heating or cooling.

Pyrolysis: A method of nanoparticle fabrication that involves decomposing a precursor material under high temperatures.

Q

Quantum Dots: Nanoparticles of a semiconducting material.

Quantum Mechanical Model of the Atom: Mathematical model of an atom that describes the dual particle-like and wave-like behavior of electrons.

R

Reactive Ion Etching (RIE): In nanofabrication, a process that uses high energy ions to remove material from a substrate's surface.

Refraction: The change in speed of light as it passes from one material to another.

Refractive Index: A number that defines how light passes through a material.

Reinforced Plastic: Composite material wherein a polymer serves as the main component with fragments added for strength.

Reinforcement: Smaller fragments of material added to a matrix which gives strength to a composite.

Resistor: A device that resists how much electrical current can flow through a system.

Secondary Batteries: Rechargeable batteries.

Secondary Bond: Forces that exist between molecules.

Self-Assembled Monolayer: A spontaneously formed nanostructure that is one molecule thick.

Self-Assembly: A process in which a solution of molecules spontaneously organize themselves into a nanostructure.

Semiconductor: A material whose electrical conductivity lies between that of an electrical conductor and that of an electrical insulator.

Semimetal: Element that has both metallic and non-metallic properties.

Sensor: A component that can measure physical quantities such as speed and convert them into a readable signal.

Significant Figures: All of the digits in a measured number that are certain.

Silica: Silicon dioxide (SiO_2).

Silicon-Controlled Rectifier: Silicon switches that control current flow.

Sol-Gel Synthesis: A method of nanoparticle fabrication that involves conversion of molecules in a colloidal solution (the 'sol') to act as a precursor for a network of nanoparticles (the 'gel')

Solar Panel: A combination of multiple photovoltaic cells capable of producing larger amounts of electricity.

Solar Power: Electricity generated by conversion of incoming solar energy via photovoltaic cells.

Solid: The state of matter characterized by rigidity, incompressibility, and having a fixed volume and shape.

Solid Lipid Nanoparticle: Nanoparticles composed of physiological lipid molecules which can carry therapeutic agents.

Spectroscopy: Any techniques that uses light to provide information about the chemical composition of a material.

Spin Coating: A process by which films are deposited by adding an excess amount of liquid material to a substrate and spinning it quickly.

Structure: Arrangement of individual atoms within a material.

Sublimation: A phase transition wherein a material changes from the solid state directly to the gaseous state.

Substrate: In materials science, a solid material onto which any process in conducted

Surface Area: Total area of the faces and curved surfaces of a solid.

Surface Area-To-Volume Ratio: The amount of surface area per unit volume of a solid object.

Surface Plasmons: Coherent oscillations between free electrons and electric fields of rays of light originating near a material.

Superalloy: A mixture of metals, generally with a nickel or cobalt base, that exhibit high mechanical strength, good surface stability and resistance to corrosion and oxidation at high temperatures.

Superhydrophobic: Property of a material that yields a very large contact angle when it comes into contact with water.

Superparamagnetism: A form of magnetism which appears in iron oxide nanoparticles wherein each nanoparticle has a fixed magnetic moment.

Systeme Internationale d'Unites (SI Units): Modern form of the metric system which comprises a universal system of units of measurement.

T

Targeted Drug Delivery: A method of delivering medicine only to localized areas where it is needed.

Tensile Strength: Maximum stress experienced by a material during a test in which it is being pulled in tension.

Tension Test: A test wherein a material is pulled in opposite directions along one axis to see how it will deform.

Thermal Cell Battery: Battery wherein the electrolyte is made up of a molten salt.

Thermal Conductivity: A material's ability to transport heat from areas of high temperature to low temperature regions.

Thermal Expansion: A material's ability to expand in volume and length upon heating.

Thermal Stress: A stress induced in a material that results from changes in temperature.

Thermopile: Temperature-sensitive MEMS devices.

Thermoplastic: Polymers that become moldable when warmed to a critical temperature.

Thermoset Plastic: Polymers that undergo an irreversible chemical change upon heating.

Titania: Titanium dioxide (TiO_2).

Top-Down Nanofabrication: A process by which nanostructures are formed into bulk material.

Toughness: Measure of the amount of energy required to fracture a material.

Transistor: A semiconducting device used to switch electronic signals.

Transmission Electron Microscope: A microscope that passes a high energy beam through a sample to provide magnified views of a material.

Travelling Lattice Waves: Coordinated vibrations among atoms that are produced when a solid is heated.

Triboelectric: Materials that generate an electric potential when rubbed together.

Triple Point: The point on a phase diagram where all three phases of matter are in equilibrium.

Tumor: A population of rapidly growing and dividing cancer cells.

U

Ultracapacitor: New type of capacitor that holds charge via an electrochemical reaction.

V

Valence Bond Theory: A theory that states a covalent bond forms when two atoms get very close to one another and one occupied orbital overlaps another orbital of a different atom.

Valence Shell: Outermost electron shell of an atom

Van der Waals Forces: Weak secondary bonds that result from temporary electron imbalances.

Vapor Condensation: A method for nanoparticle synthesis that involves formation when metallic vapors are quickly condensed.

Vaporization: A phase transition wherein a material changes from the liquid state to the gaseous state.

W

Water Pollution: Contaminants in water that do not support human use or impair water's ability to support biotic communities.

Wave: A continuously repeating change or oscillation in matter or a physical field.

Wavelength: The distance between any two adjacent, identical parts on a wave.

Wet Cell Battery: A battery wherein the electrolyte material is in liquid form.

Wet Etching: In nanofabrication, a process that uses wet liquid chemicals to remove material from a substrate's surface.

X

X-Ray Photoelectron Spectroscopy: A technique that measures elemental composition of the top 1-10 nm of a thin film's surface.

X-Ray Spectroscopy: Any method that uses characteristic X-rays produced by a material when it interacts with high energy electrons.

Printed in the United States
By Bookmasters